西南山地SWILD 编

中国西南山地
珍稀特有鸟类

Rare and Endemic Birds in
Mountains of Southwest China

四川美术出版社

　　西南山地SWILD秉承科学精神,在大自然中创作,构建中国自然影像库,通过生物多样性多媒介调查监测评估、自然记录宣教片制作、自然书籍画册出版等方式,诠释社会公众和自然保护之间的和谐与共生,是知名的生物多样性影像科学传播平台。

《中国西南山地珍稀特有鸟类》编委会

主　　任：曾　雨　杨贵伟（高黎贡山国家级自然保护区怒江管护局）
　　　　　唐　军　罗平钊　唐海涛
副 主 任：董　磊　何　屹　徐　健
委　　员：朱　磊　关翔宇　巫嘉伟　杨小农　李　茂　刘　丽
　　　　　孟姗姗　何既白　周　天　胡　佳　路家兴　张秀雷
　　　　　于鹏飞　李　祺　关卫东　王　芳　谢奇言　谭　琦
　　　　　董晏薇　袁一帆　林雪红　管丽琼　丹曾央吉　邓巴腊姆

总 策 划：西南山地SWILD
内容统筹：刘　丽　孟姗姗　何既白　周　天
助理编辑：胡　佳　路家兴　张秀雷　于鹏飞　李　祺　关卫东
　　　　　王　芳　谢奇言　谭　琦
撰　　文：朱　磊　唐　军　关翔宇　巫嘉伟
绘　　画：杨小农
英文翻译：亦趣自然

目录 CONTENTS

| 8 | 序言 Foreword |
| 10 | 概况 Overview |

鸟类 Birds

- 004 1. 青头潜鸭 Baer's Pochard
- 008 2. 中华秋沙鸭 Scaly-sided Merganser
- 010 3. 斑尾榛鸡 Chinese Grouse
- 012 4. 红喉雉鹑 Chestnut-throated Monal Partridge
- 014 5. 黄喉雉鹑 Buff-throated Monal Partridge
- 016 6. 四川山鹧鸪 Sichuan Partridge
- 018 7. 灰胸竹鸡 Chinese Bamboo Partridge
- 020 8. 红腹角雉 Temminck's Tragopan
- 022 9. 白尾梢虹雉 Sclater's Monal
- 026 10. 绿尾虹雉 Chinese Monal
- 028 11. 白马鸡 White Eared Pheasant
- 030 12. 蓝马鸡 Blue Eared Pheasant
- 032 13. 红腹锦鸡 Golden Pheasant
- 034 14. 白腹锦鸡 Lady Amherst's Pheasant
- 038 15. 黑颈鹤 Black-necked Crane
- 040 16. 林沙锥 Wood Snipe
- 042 17. 大紫胸鹦鹉 Derbyan Parakeet
- 044 18. 鹊鹂 Silver Oriole
- 046 19. 黑头噪鸦 Sichuan Jay
- 048 20. 红腹山雀 Rusty-breasted Tit
- 050 21. 四川褐头山雀 Sichuan Tit
- 052 22. 领雀嘴鹎 Collared Finchbill
- 054 23. 银脸长尾山雀 Sooty Bushtit
- 056 24. 峨眉柳莺 Emei Leaf Warbler
- 058 25. 四川短翅蝗莺 Sichuan Bush Warbler
- 060 26. 金额雀鹛 Golden-fronted Fulvetta
- 062 27. 红翅噪鹛 Red-winged Laughingthrush
- 064 28. 灰头斑翅鹛 Streaked Barwing
- 066 29. 橙翅噪鹛 Elliot's Laughingthrush
- 068 30. 灰胸薮鹛 Emei Shan Liocichla

070	31. 画眉 Hwamei	110	50. 白脸鸦 Przevalski's Nuthatch
072	32. 黑额山噪鹛 Snowy-cheeked Laughingthrush	112	51. 巨鸦 Giant Nuthatch
074	33. 大噪鹛 Giant Laughingthrush	114	52. 四川旋木雀 Sichuan Treecreeper
076	34. 白点噪鹛 White-speckled Laughingthrush	116	53. 棕背黑头鸫 Kessler's Thrush
078	35. 斑背噪鹛 Barred Laughingthrush	118	54. 宝兴歌鸫 Chinese Thrush
080	36. 棕草鹛 Tibetan Babax	120	55. 棕头歌鸲 Rufous-headed Robin
082	37. 棕噪鹛 Buffy Laughingthrush	124	56. 黑喉歌鸲 Blackthroat
084	38. 火尾绿鹛 Fire-tailed Myzornis	126	57. 金胸歌鸲 Firethroat
088	39. 宝兴鹛雀 Rufous-tailed Babbler	128	58. 栗背岩鹨 Maroon-backed Accentor
090	40. 中华雀鹛 Chinese Fulvetta	130	59. 拟大朱雀 Streaked Rosefinch
092	41. 棕头雀鹛 Spectacled Fulvetta	132	60. 曙红朱雀 Pink-rumped Rosefinch
094	42. 灰头雀鹛 Grey-hooded Fulvetta	134	61. 淡腹点翅朱雀 Sharpe's Rosefinch
096	43. 三趾鸦雀 Three-toed Parrotbill	136	62. 斑翅朱雀 Three-banded Rosefinch
098	44. 白眶鸦雀 Spectacled Parrotbill	138	63. 白眉朱雀 Chinese White-browed Rosefinch
100	45. 暗色鸦雀 Grey-hooded Parrotbill	140	64. 蓝鹀 Slaty Bunting
102	46. 灰冠鸦雀 Rusty-throated Parrotbill		
104	47. 黄额鸦雀 Fulvous Parrotbill	142	附录一 西南山地观鸟线路攻略
106	48. 白领凤鹛 White-collared Yuhina	153	附录二 索引
108	49. 滇鸦 Yunnan Nuthatch	157	附录三 山地六条

序 言
Foreword

中国西南山地是一个地理名词，有人认为即是横断山区，也有人认为是横断山区的一个西南山地亚区，尚无统一说法。且西南山地是由英文Mountains of Southwest China翻译而来，可泛指生物多样性丰富的中国西南山地地区。

中国西南山地是非常具有代表性的地理概念，涉及中国一成的国土面积[1]，区域内主要为山地和湿地，其中著名的大熊猫、绿尾虹雉等物种特属于此，是国际社会公认的全球36个生物多样性热点区域之一[2]。自2004年起，成都的生态文化团队以"西南山地"作为自己的团队名称，树立了业内的生态品牌形象，并建立了原创影像网站www.swild.cn。今日，"西南山地"更成为一类热爱中国西南地区生物多样性的人群象征。

据统计，中国鸟类特有种模式产地主要分布于17个省份，其中尤以西南的四川最多，达27种和亚种，国外人士命名的年代主要集中在1860年至1880年之间，而中国学者命名的年代主要在20世纪70年代以后。

您面前的这本书历时六年而成，我们希望汇集符合"山地六条"（见附录三）自然摄影行业标准的野鸟摄影作品来展示真实鲜活的科学信息和飞羽灵动的珍稀鸟类。

这是一本面向观鸟者和鸟类拍摄者的书，希望站在国际视野上，精选中国，主要是西南山地记录的珍稀、特有或具有全球关注度的特别鸟类，按此逻辑对珍稀特有鸟类进行选择，除了参考《国家重点保护野生动物名录》，更重点关注世界自然保护联盟濒危物种红色名录中"易危"以上级别的鸟类，由此精编而成本书。

---

[1] CEPF（关键生态系统合作基金）：《中国西南山地生物多样性热点地区》，官网www.cepf.net，"该热点地区约占中国地理面积的10%"。
[2] 根据CEPF的统计，截至2023年8月，国际社会公认的生物多样性热点地区有36个。

灰胸薮鹛　摄影：叶昌云

概 况
Overview

朱 磊

鸟类生态学和保护生物学研究工作者
生态学博士
译者

1988年，英国学者诺曼·迈尔斯（Norman Myers）在对热带雨林受威胁程度进行分析的时候，依据特有植物种类的丰富程度和栖息地所面临丧失的威胁，提出了"热点地区"（hotspot）这一概念。经过十余年的不断完善，并且在汲取了相关领域研究成果的基础之上，他与合作者于2000年将这一概念应用到了全球，根据不同地区特有植物及特有陆生脊椎动物及其受威胁程度，评估出世界25个生物多样性热点地区。入选生物多样性热点地区必须满足两个条件：第一，该地区至少应包括1500种特有的维管植物（约占世界植物种类总数的5‰）；第二，该地区已经丧失了不少于70%的原始自然栖息地。"热点地区"这一概念能够比较好地应用于确定不同地区的保护优先序，也能很好地用以评估保护工作的有效性，一经推出便很快得到了广泛的关注及认可。截至2023年8月，全球已有36个生物多样性热点地区被甄别出来，它们总共只占到地球陆地面积的2.3%，但却包含了近50%的特有植物和42%的陆生脊椎动物种类。这些区域在具有极高保护价值的同时，也正面临着栖息地丧失的严重威胁。

无论是早年的25个，还是如今的36个的名单中，都收录了一个叫作"中国西南山地"（Mountains of Southwest China）的生物多样性热点地区。该区域在我国西藏东部、四川西部及云南西部一带，为岷江和雅鲁藏布江南北向河谷之间的一系列南北走向山脉，也被习惯称作"横断山区"。这里地处青藏高原的东南缘，山川并列，峰高谷深，在地貌上是高原的延续，但地质构造相较年轻的高原却又显得古老而复杂。区内自然地理条件独特，生物区系组成复杂而多样，既包含如大熊猫（$Ailuropoda\ melanoleuca$）这样有着三四百万年历史的相对古老成分，又是显而易见的许多鸟类的现代分布中心。从动物地理学的角度来看，横断山区处在东洋界西南区西南山地亚区与古北界青藏区青海藏南亚区交汇的地带，在海拔较高的地方，有不少古北界的种类沿山脊向南伸展分布，而不少属于东洋界的热带种类则沿海拔较低的河谷向北分布，两界成分混杂且呈现垂直变化，水平分界线则不甚明显。同时，现有的研究还提示横断山区在严酷的冰期曾成为避难所，不少古老的物种得以保留于此。这里也因此成为举世瞩目的全球生物多样性热点地区之一。

地理与地貌

我国的山脉大多呈东西走向，但西南山地的主要山脉则呈近南北走向。

提及西南山地中"横断山脉"一词，已知的最早出处，是邹代钧编写的《京师大学堂中国地理讲义》（1900—1901），其中便有"迤南为岷山，为雪岭，为云岭，皆成自北而南之山脉，是谓横断山脉"的记载。但邹先生虽用了"横断山脉"的说法，却并未进一步解释缘由。其后，有观点认为这一名称源自西方学者，但经陈富斌（1984）[①]考证，国外学者对川滇藏一带的一系列南北走向山脉并无统一称谓。严德一（1942）[②]认为"盖因南北行列之高山深谷，阻碍人类东西交通之动向，是以'横断'名之也"。此观点为之后的一些著作引述，颇具一定的影响。但陈富斌（1984）指出我国许多东西向的山脉，如秦岭、天山和唐古拉山等，历史上一样产生了阻隔交通的影响，但并未以"横断"相称。"横断"一词的真正含义或许已不可考，但用于指称这一系列的南北走向山脉却已是约定俗成的事实。

人们不仅对如何解释"横断"有着不同看法，对于"横断山区"所指代的具体范围也存在各种观点。关于具体东、西、南、北四条边界的认定，过去也没有一致的意见，由此也衍生出了"三脉说"（三山夹两江）、"四脉说"（四山夹三江）、"五脉说"、"六脉说"，乃至"七脉说"，实际也就对应了所谓"广义"或"狭义"上的横断山区范围。李炳元（1987）[③]在经过仔细研究之后，支持"七脉说"，亦即"广义"的横断山区范围：西起伯舒拉岭—高黎贡山，东至岷山，北抵昌都—甘孜—马尔康一线，南到云南龙陵—南涧—下关—丽江—四川盐源一线。区内自西向东有着七列近乎平行的南北向山脉：伯舒拉岭—高黎贡山、他念他翁山—怒山、芒康山—云岭、沙鲁里山、大雪山、邛崃山和岷山。

由此一来，横断山区北接秦岭、阿尼玛卿山和巴颜喀拉山，西部与唐古拉山和念青唐古拉山相连，南面则同滇西南和滇中南山原相邻，东向则是秦岭—大巴山、四川盆地和川西南山地。行政区划上主要跨川、滇、藏三省区，包括四川省甘孜藏族自治州、阿坝藏族羌族自治州、凉山彝族自治州、雅安市部分地区和攀枝花市，云南省迪庆藏族自治州、丽江市、大理白族自治州、怒江傈僳族自治州、楚雄彝族自治州和保山市部分地区，西藏自治区昌都地区大部，以及青海和甘肃两省的极南部。区内总面积约36.4万km²，其中92%以上属青藏高原，为高原东南部的重要组成部分（李炳元，1987；李文华等，2010[④]）。上述范围，即是本书所指的"横断山区"。

西南山地地势为明显的北高南低，由南向北海拔逐渐升高，大致可分为三级阶梯：第一级自云南泸水至四川普格以南，平均海拔约在2000m以下，但也有个别山峰较高，如高黎贡山最高达3374m；第二级自云南德钦至四川康定以南，平均海拔约2700m，呈现极高山和深切河谷地貌，其中碧罗雪山4379m，玉龙雪山最高

[①] 陈富斌：《"横断山脉"一词的由来》，载于《山地研究》，1984年3月第2卷第1期，第31—35页。
[②] 严德一：《横断山脉中之气候蠡测》，载于《气象学报》，1942年，第140—150页。
[③] 李炳元：《横断山脉范围探讨》，载于《山地研究》，1987年6月第5卷第2期，第74—82页。
[④] 李文华、张谊光主编《横断山区的垂直气候及其对森林分布的影响》，气象出版社，2010年。

摄影：左凌仁

峰5596m，哈巴雪山5396m，梅里雪山主峰卡瓦格博峰达6740m；第三级为德钦至康定以北的丘状高原，平均海拔已升至4000m以上，山峰高度则均超过6000m，贡嘎山主峰高达7556m，不仅是四川境内的最高峰，也是整个西南山地的最高点（唐蟾珠，1996[①]；王襄平等，2004[②]）。

区内的各大山脉还呈现自北向南扇状展开之势，地形基干骨架由自西向东的五大山系构成（唐蟾珠，1996；李文华等，2010）：

1. 伯舒拉岭—高黎贡山　东西走向的念青唐古拉山在97°E附近转为了南北延伸，开始被称作伯舒拉岭，再向南进入云南之后，即称高黎贡山。伯舒拉岭主峰查格腊子峰海拔6155m，而高黎贡山主峰则仅有3374m。

2. 他念他翁山—怒山—碧罗雪山　唐古拉山向东延伸至西藏东部称他念他翁山，进入云南西北部后称作怒山，再南延则称碧罗雪山。怒山以东为流入太平洋的澜沧江水系，以西则是注入印度洋的怒江水系。滇藏交界处梅里雪山海拔6740m的主峰卡瓦格博峰是西南山地第二高峰，而从卡瓦格博峰至澜沧江边的西当铁索桥（海拔1980m），直线距离不过12km，相对高差则达4760m之巨。

3. 芒康山—云岭—无量山[③]　芒康山也是唐古拉山的分支，在西藏昌都地区称为芒康山，至云南境内则称云岭。再向南又分为两支，雪盘山和清水朗山，继续南延所成的余脉即是无量山。芒康山—云岭是澜沧江和金沙江的分水岭，该山系内海拔5000m以上的山峰即有20余座，其中白马雪山主峰扎拉雀尼峰海拔5640m。

4. 沙鲁里山—玉龙雪山—哀牢山　这一山系为金沙江和雅砻江的分水岭，西北起自雀儿山，向东至四川甘孜分为素龙山，顺雅砻江向南延至理塘后称九拐山，至稻城以南即为贡嘎雪山。向西的一支在甘孜境内往南称沙鲁里山，进入云南之后东西两支又交汇成玉龙雪山。玉龙雪山再向南延，形成呈西北—东南走向的哀牢山。玉龙雪山的主峰扇子陡海拔5596m，这里发育的冰川是我国纬度最低的一座现代冰川。

5. 牟尼芒起山—大雪山—锦屏山　这一山系为雅砻江与大渡河的分水岭，山脊海拔普遍在5000m以上。四川道孚以南称大雪山，主峰贡嘎山海拔7556m，被誉为"蜀山之王"。贡嘎山东坡下部的大渡河河谷海拔仅1100m，距主峰直线距离29km，但相对高差超过了6400m。大雪山向南延伸至雅砻江大拐弯以南称作锦屏山，主峰海拔约4080m。

在川滇藏交界不到60km宽的范围内，有南北纵贯

[①] 唐蟾珠主编《横断山区鸟类》，科学出版社，1996年。
[②] 王襄平、王志恒、方精云：《中国的主要山脉和山峰》，载于《生物多样性》，2004年第12卷第1期，第206—212页。

[③] CEPF认为无量山和哀牢山不属于横断山区，而应归入印缅热点（Indo-Burma Hotspot）。

的大雪山、云岭、怒山、高黎贡山四列山脉和金沙江、澜沧江、怒江三条大河相间并行的奇特景观，即著名的"三江并流"。这一"江水并流却不交汇"的壮美独特的自然地理景观，已于2003年被收入世界自然遗产名录。

总体而言，山川相间排列，岭谷之间地势落差悬殊是区内最为显著的地貌特征。由于所流经地区在地质构造、地势等方面存在差异，区内河流切割而成的河谷也宽窄不一。如金沙江上著名的虎跳峡峡谷宽仅60~80m，而金沙江的一级支流雅砻江的上游河谷却可宽达1000~2000m。金沙江、澜沧江和怒江在30°N以南切割的深度往往可达1000~2000m，形成典型的"V"字形深切河谷。除开这些海拔变化剧烈的岭谷，在本区中还有许多大小不一、成因不同的盆地，著名的断陷盆地如大理、元谋、西昌等，今天都是人口聚集的地方。1965年开始相继在元谋盆地发现了元谋直立人（*Homoerectus yuanmouensis*）、蝴蝶禄丰古猿（*Lufengpithecus hudienensis*）的化石，表明早期人类也曾活跃在这一区域之中。而丽江、中甸（今香格里拉）和泸沽湖这样的喀斯特盆地则早以秀丽的山水风光闻名于世。

气候与植被

西南山地主要受到西南季风和高原季风的影响，呈现明显的干湿季。每年11月至次年4月在高空西风急流的控制下，空气干燥，降水少而气温低，形成干季。而5至10月随着西风带北移，在西南季风的影响下，空气湿度增大，云雨天气开始增多。与此同时，高原夏季风的建立又进一步增强了西南季风，随着气温的升高，降水也逐渐增加，这一时期的降水量可达全年的80%~95%，是名副其实的雨季。北上的暖湿气流依山势从东南、西南方向逐渐往西南山地中部深入，雨季也自区内的东南、西南部向中部依次发展。西南部的云南怒江州北部受气象和特殊地形条件共同作用，3至4月间即进入雨季，是西南山地乃至全国雨季开始最早的地方。而在贡嘎山以东，受东南暖湿气流影响，雨季始于4月中下旬。区内其他大部分地方雨季多于5月左右开始，而中部的河谷地带因水汽受到两侧高大山体的阻挡，要迟至6月上中旬才进入雨季。进入10月，多数地方的雨季即行结束（李文华等，2010）。

区内气候还受到纬度和经度的显著影响，表现为纬度地带性更强，经向变化更为显著。从南到北，从低到高，按照热量状况的差异，气候带依次呈现热带、亚热带、温带至寒带的变化。而东西方向上，既有山地湿润、半湿润，也有河谷半干旱、高原半干旱的气候类型。由于区内地势由南向北逐渐升高，从而加剧了因纬度升高而产生的气候差异。气温从南到北逐渐降低，例如同在99°E，南部的六库年平均气温20.2℃，处于北部的石渠则为-1.6℃，一南一北间年平均气温差距可达

21.8℃。两地之间的纬度相差约为8°，即纬度北移1°，年平均气温降低约3℃（李文华等，2010）。

区内的降水量从东西两侧向中部呈波浪状变化，其中西侧因受印度洋西南季风深入的影响，降水量要大于东侧。东西两侧又大于中部，山地则多于河谷。东南暖湿气流受岷山山脉、邛崃山山脉等山体阻挡，在四川盆地的西缘形成了著名的"华西雨屏带"，是我国内陆地区降水最为丰沛的区域。以四川雅安为例，年降水日数可多达263.5天，年降水量超过1700mm。而在雅安境内夹金山脉海拔3437m的二郎山，当地民谚便有"二郎山顶分世界，西边日头东边雨"的说法。来自四川盆地的潮湿气流，在华西雨屏一侧的迎风坡抬升，随着温度降低，水汽发生凝结，最终形成充沛的降雨。等气流翻过山脊，下坡过程中水汽蒸发，云量减少，河谷地带常常阳光普照。背风坡的年降水日数变为约145天，年降水量也下降至640mm。降水量不仅存在东西方向上的变化，从南到北也存在明显差异。暖湿气流沿河谷北上，其影响也在逐渐减小。从最南的大渡河河谷康定，依次经雅砻江河谷雅江、金沙江河谷巴塘、澜沧江河谷芒康至最北的怒江河谷八宿，年降水量也从1700mm、700~800mm、400~500mm，减为260mm（林之光，1995[1]；唐蟾珠，1996；庄平等，2002[2]；李文华等，2010）。

降水分布的不均匀性，在西南山地当中形成了非常独特的"干旱河谷"现象，也是我国湿润、半湿润地区当中存在的特殊自然地理类型。在梅里雪山东侧的澜沧江河谷及白马雪山东侧的金沙江河谷表现尤为明显，金沙江河谷的德钦奔子栏北至甘孜得荣一带年降水量仅约300mm，和内蒙古西部、宁夏、甘肃等干旱地区相仿，也因此被称为"西南干旱中心"。这些在深切河谷中形成的彼此隔离的局部性半干旱—干旱气候，根据热量的差异，又可分为干热、干暖和干温河谷气候，是典型的非地带性气候（林之光，1995；唐蟾珠，1996；刘晔等，2016[3]）。

最后，垂直变化剧烈的山地气候是西南山地的鲜明特征。以区内东北部的四川盆地西缘向青藏高原过渡的卧龙—四姑娘山为例，东坡坡度较为平缓，坡下部年平均气温垂直递减率为0.46℃/100m，坡上部则是0.53℃/100m；年降水量随海拔升高而增加，坡下部河谷约为850mm，巴朗山垭口为1380mm。西坡地势较陡，坡下部年平均气温垂直递减率为0.77℃/100m，坡上部则是0.75℃/100m，明显大于东坡；年降水量虽也随海拔升高而增加，但少于东坡，坡下部河谷约为600mm，坡上部约为800mm。因此东坡仍属四川盆地周边的半湿润气候，西

---

[1] 林之光：《地形降水气候学》，科学出版社，1995年。
[2] 庄平、高贤明：《华西雨屏带及其对我国生物多样性保育的意义》，载于《生物多样性》，2002年第10卷第3期，第339—344页。
[3] 刘晔、李鹏、许玥等：《中国西南干旱河谷植物群落的数量分类和排序分析》，载于《生物多样性》，2016年第24卷第4期，第378—388页。

坡则已变为高原河谷半干旱气候（李文华等，2010）。

植物被认为是反映气候的一面镜子。一个地方有什么样的气候，通常就会对应生长什么样的植物。西南山地山高谷深，地形地势复杂，气候变化剧烈，从河谷到山地，也孕育了多样的植被类型。区内山地垂直自然带变化极为明显是一大特征，从山麓到山顶往往具有边缘热带、亚热带、山地暖温带至高原亚寒带等各种植被类型，是我国乃至世界高山植被带最为丰富的地区之一（唐蟾珠，1996；李文华等，2010；刘晔等，2016）。大致可分为7类：

1. 热带北缘山地雨林、季雨林 分布于高黎贡山西南角和怒江、金沙江深切河谷盆地海拔1000~1200m区域。高黎贡山南段迎风坡受西南季风影响，气候湿热，雨量充沛，年降水量1500~2000mm，森林主要由无患子科、梧桐科、使君子科等种类构成，但原始林多已遭破坏。河谷盆地的落叶乔木主要由桑科、木棉科、四数木科等种类组成。

2. 山地亚热带常绿阔叶林 该类型是区内南段山地的基本林型，一般分布于海拔1600~1800m区域，随着海拔的不同，植物种类组成略有差异。海拔较低处多由喜温的亚热带树种和部分热带树种组成，乔木多是壳斗科、樟科、山茶科等种类。海拔较高处则是常绿与落叶林，在有些地方海拔可至2700m，多为樟、楠等阔叶混交林，青冈阔叶混交林等。

3. 硬叶常绿栎林 该林型在世界范围内主要分布在欧洲地中海、北美太平洋沿岸及我国西南山地，是以季风气候为特征的一个较为特殊的森林类型。其中高山栎类分布海拔上限一般为2600~3500m，个别地方能达到4000m。暖性灌木高山栎通常见于海拔2300~2600m的山地亚热带和山地温带谷地斜坡，而温性川滇高山栎则在西藏昌都及邻近地区海拔2800~3900m的山地阳坡、半阳坡较为常见。松萝垂挂的川滇高山栎则生长于亚高山地带温凉湿润的环境中，主要分布在西藏东南及周边海拔3500m以上的山地阴坡和半阳坡。

4. 松林 区内的松林主要为华山松林、高山松林和云南松林三大类。华山松是我国特有针叶树种，主要分布在西南山地南段，常参与形成针阔混交林。高山松则是区内针叶林的主要建群树种，分布范围东起川西岷江流域，西至西藏中部，南至云南西北，北抵川西道孚南部，常见于海拔3500m以下的山地阳坡。云南松对干旱适应力较强，常见于23°~29°N，96°~106°E之间的广袤区域，多分布在海拔1500~3000m的区域。

5. 暗针叶林 北半球分布广泛的一种林型，是亚高山地带的典型植被。区内多由云杉属、冷杉属和铁杉属的一些种类构成。云杉林在怒江峡谷以东广大地区的分布海拔上限多为3100~3900m，以丽江云杉、川西云杉、林芝云杉等为建群种。冷杉林则以海拔3500m以上最为集中，最高可达4300m左右，建群种为川滇冷杉、西藏冷杉、长苞冷杉等，依据林下混合植被的组成又可分为灌木草类冷杉林、箭竹冷杉林、苔藓冷杉林和杜鹃

摄影：邹滔

冷杉林四大类。

6. 高山灌丛草甸　森林向高海拔草甸过渡的类型，以杜鹃灌丛、高山栎灌丛和高山柳灌丛为主，海拔再上升则多为蒿草草甸。

7. 干旱河谷植被　区内干旱河谷中耐旱的稀树灌丛或灌（禾）草丛，又可划分为半萨王纳植被（semi-savanna）和河谷型马基植被（Maquis）。

总体而言，这里的植物在相对封闭而又独具特色的气候条件下经历了长期的适应与演化过程，既保存了较高的物种多样性，又包括较多的古老孑遗成分。

鸟类概览

如前所述，西南山地有着崎岖陡峭的地形、多种多样的气候条件和与之对应的植被类型，使得这里的高山深谷成为种类繁多的动物的家园。而那些高耸入云的山峰，恰如一个个白云之上的岛屿，与周围环境相对隔绝的状态，为其间生物的独立演化创造了必要条件，也是西南山地兼具很高的生物多样性和物种特有性的重要原因。区内面积仅占中国陆地总面积的10%，但却拥有超过一半的中国鸟类和哺乳动物种类，其中不乏众多珍稀濒危物种。该区域被认为是雉类、噪鹛类和鸦雀类等的分布中心及可能的起源地，也保有了全球生物多样性最为丰富的温带森林生态系统。

这一区域究竟有多少种鸟类？对于这个问题，过去不同的研究者都给出了自己的答案。唐蟾珠（1996）依据野外采集的标本、实际观察和文献记载认为区内分布鸟类585种，约占当时已知中国鸟类种数的46.7%（郑作新，2000）[1]。Cristina Mittermeier等（2004）指出区内有记录鸟类611种，占已知总数的45.9%（郑光美，2005）[2]。Y Wu等（2013）[3]则根据馆藏标本、观察记录和文献记载整理，认为这一区域繁殖鸟类为738种，占总数的51.1%（郑光美，2017）[4]。[5]由于上述研究的年代不同，所涉及的具体地理范围也不一致，在鸟类种数上自然也存在差异。目前尚无对于西南山地鸟类种数的最新权威统计，但根据已有数据保守估计，区内分布鸟类应约占全国已知总数的一半，是当之无愧的生物多样性热点地区。

西南山地地形复杂，气候多变，孕育了多样化的植

---

[1] 郑作新：《中国鸟类种和亚种分类名录大全》，科学出版社，2000年。

[2] 郑光美：《中国鸟类分类与分布名录》，科学出版社，2005年。

[3] Y Wu, RK Colwell, C Rohbek, etc., "Explaining the Species richness of birds along a subtropical elevational gradient in the Hengduan Mountains," Journal of Biogeography, 2013,40(12),p.2310–2323.

[4] 郑光美：《中国鸟类分类与分布名录（第三版）》，科学出版社，2017年。

[5] 须指出的是，各家对于横断山区所涉面积大小说法并不统一，本书采信区域总面积约36.4万km²的观点（李炳元，1994；李文华等，2010）；而唐蟾珠（1996）则认为总面积约50万km²；Cristina Mittermeier等（2004）所认可的面积最小，仅约26.2万km²；Y Wu等（2013）统计的范围最大，面积达66万km²。所涉面积的不同，加之近来鸟类分类上的发展，必然会导致研究中最终统计到的鸟类种数上的差异，这一点值得引起注意。

被类型，为生活在其中种类繁多的鸟类提供了各种栖息环境。不同的生境中分布着不同类群的鸟类，以下简述区内大致的生境类型和其中的代表鸟种（唐蟾珠，1996）：

1. **干热河谷灌丛带** 气候干燥少雨；植被以耐旱灌丛为主；鸟类则以东洋界成分为主，特有种、古北界种类和非繁殖鸟均较少。代表性鸟种有棕胸佛法僧（*Coracias benghalensis*）、灰燕鵙（*Artamus fuscus*）、红耳鹎（*Pycnonotus jocosus*）和蜂虎（*Merops* spp.）等。

2. **亚热带低山常绿阔叶林带** 气候炎热湿润；植被以高大乔木为主；鸟类则以东洋界成分为主，古北界种类和非繁殖鸟较少。代表鸟种有黑颈长尾雉（*Syrmaticus humiae*）、褐林鸮（*Strix leptogrammica*）、针尾绿鸠（*Treron apicauda*）和斑翅鹛（*Actinodura* spp.）等。

3. **暖温带中山针阔混交林带** 气候温凉湿润；植被以高大乔木为主；鸟类中东洋界成分优势已不明显，特有种比例上升，古北界种类和非繁殖鸟所占比例最低。代表鸟种有环颈山鹧鸪（*Arborophila torqueola*）、凤头鹰（*Accipiter trivirgatus*）、宝兴歌鸫（*Turdus mupinensis*）和黄嘴蓝鹊（*Urocissa flavirostris*）等。

4. **寒温带亚高山针叶林带** 气候依然温凉湿润，但气温普遍较针阔混交林带低；植被以高大云杉、冷杉属成员为主，林下往往着生茂密的箭竹；东洋界鸟类最少，特有种和古北界种类所占比例最高。代表鸟种有红腹角雉（*Tragopan temminckii*）、黄额鸦雀（*Suthora fulvifrons*）和黑头噪鸦（*Perisoreus internigrans*）等。

5. **亚寒带高山灌丛草甸带** 气候寒冷；植被以高寒草甸灌丛为主；古北界鸟种占绝对优势。代表鸟种有绿尾虹雉（*Lophophorus lhuysii*）、猎隼（*Falco cherrug*）、长嘴百灵（*Melanocorypha maxima*）和地山雀（*Pseudopodoces humilis*）等。

6. **河流湖泊沼泽湿地** 区内河流湖泊众多，沼泽湿地则主要见于四川甘孜州和阿坝州境内；鸟类多以迁徙经过的旅鸟和冬候鸟为主，也有在高原湿地上繁殖的种类。代表鸟种有大天鹅（*Cygnus cygnus*）、黑颈鹤（*Grus nigricollis*）、鹮嘴鹬（*Ibidorhyncha struthersii*）和红尾水鸲（*Phoenicurus fuliginosus*）等。

7. **村寨农田耕地环境** 区内中低海拔地势较平坦，水热条件较好的地方多已被人类生活和生产活动极大改变了原有面貌。在此环境中生活的鸟类以东洋界成分和非繁殖鸟为主。代表鸟种有黑翅鸢（*Elanus caeruleus*）、金腰燕（*Cecropis daurica*）、麻雀（*Passer montanus*）和凤头鹀（*Emberiza lathami*）等。

从各个生境分布鸟种数来看，亚热带低山常绿阔叶林是很多种鸟类的理想栖息地，该生境中也生活着最为丰富的鸟种；其次是暖温带中山针阔混交林；而干热河谷灌丛是区内鸟类多样性最为贫乏的生境（唐蟾珠，1996）。

上述生境在西南山地中的分布除受经纬度变化影响以外，在同一山脉当中还会随海拔变化而呈现出明显的垂直地带性，进而也影响到鸟类的垂直分布。唐蟾珠（1996）将区内的鸟类垂直分布依据海拔大致划为800~1500m、1600~2500m、2600~3500m、3600~4500m这四个垂直带进行分析。结果发现海拔1600~2500m的中低山地带鸟种数最为丰富，对应的自然植被主要为半湿润常绿阔叶林，原生植被遭受破坏后常被云南松林所取代。除山地森林之外，这一海拔段还有山间河谷盆地、高原湖泊及沼泽湿地，也为许多依附农耕地和湿地生活的鸟类提供了生活环境。而Y Wu等（2013）基于繁殖鸟类的研究则发现区内鸟种丰富度最高的海拔范围为800~1800m，认为这一海拔段内丰富的鸟种数与相对稳定而优越的气候因素和较高的生产力有关。

以上是从生境和海拔垂直分布的角度来考察生活在区内的鸟类，而在动物地理学上，西南山地也拥有属于自己的特殊地位。不同的动物类群，受其演化历史背景、外界环境条件和自身扩散能力等因素的影响，形成了不同的分布格局，或称分布型。张荣祖（2011）[①]在归纳和总结中国陆栖脊椎动物物种的分布型时，专门列出了喜马拉雅—西南山地型，指主要分布于西南山地中低山，乃至延伸至喜马拉雅山脉南坡的种类。它们多生活于山地森林，为东洋界成分。地质历史时期，随着青藏高原在距今300万年—250万年的第四纪早期的隆起，高原内部气候逐渐趋向寒冷干旱，环境向高寒荒漠及草原发展，森林被迫向高原东南边缘退缩，西南山地和喜马拉雅山区由此成为偏好暖湿环境物种的避难所，并在此过程中形成了物种分布的喜马拉雅—西南山地型。其中又可分为西南山地亚型，代表鸟种有斑尾榛鸡（*Tetrastes sewerzowi*）、宝兴鹛雀（*Moupinia poecilotis*）和中华雀鹛（*Fulvetta striaticollis*）等，这三种也均为中国特有鸟类。此外还有喜马拉雅—西南山地交汇地带亚型，代表鸟种为血雀（*Carpodacus sipahi*）、纹胸斑翅鹛（*Actinodura waldeni*）和白颈凤鹛（*Yuhina bakeri*）等。

而在动物地理区划上，西南山地被划为东洋界中印亚界西南区西南山地亚区。由于区内独特的南北走向高山岭谷，形成了南边的东洋界和北边的古北界动物交流的天然通道。海拔较高处，不少古北界种类沿山脊向南伸展，而不少热带、亚热带种类则可沿河谷向北分布，由此也造就了动物区系中南北成分混杂的现象（张荣祖，2011）。唐蟾珠（1996）进一步指出古北界和东洋界在西南山地交汇界线的变化规律，即纬度越低交汇界线的海拔越高，反映出纬度越低，东洋界成分的垂直分布海拔上限越高，反之亦然。

除南北成分混杂以外，西南山地鸟类的另一特点是特有种丰富，除前述绿尾虹雉（*Lophophorus lhuysii*）、斑尾榛鸡（*Tetrastes sewerzowi*）、黑头噪鸦（*Perisoreus internigrans*）、宝兴歌鸫（*Turdus*

[①] 张荣祖：《对中国动物地理学研究的几点思考》，载于《兽类学报》，2011年第31卷第1期，第5—9页。

mupinensis)、宝兴鹛雀（*Moupinia poecilotis*）和中华雀鹛（*Fulvetta striaticollis*）之外，还有红喉雉鹑（*Tetraophasis obscurus*）、黄喉雉鹑（*T. szechenyii*）、灰胸竹鸡（*Bambusicola thoracicus*）、红腹锦鸡（*Chrysolophus pictus*）、白马鸡（*Crossoptilon crossoptilon*）、蓝马鸡（*C. auritum*）、四川林鸮（*Strix davidi*）、白眉山雀（*Poecile superciliosus*）、红腹山雀（*P. davidi*）、四川褐头山雀（*P. weigoldicus*）、四川短翅蝗莺（*Locustella chengi*）、甘肃柳莺（*Phylloscopus kansuensis*）、银脸长尾山雀（*Aegithalos fuliginosus*）、银喉长尾山雀（*A. glaucogularis*）、凤头雀莺（*Leptopoecile elegans*）、褐头雀鹛（*Fulvetta manipurensis*）、三趾鸦雀（*Cholornis paradoxus*）、白眶鸦雀（*Sinosuthora conspicillata*）、暗色鸦雀（*S. zappeyi*）、灰冠鸦雀（*S. przewalskii*）、褐顶雀鹛（*Schoeniparus brunneus*）、金额雀鹛（*S. variegaticeps*）、斑背噪鹛（*Ianthocincla lunulata*）、白点噪鹛（*I. bieti*）、大噪鹛（*I. maximus*）、黑额山噪鹛（*I. sukatschewi*）、山噪鹛（*Pterorhinus. davidi*）、棕噪鹛（*P. berthemyi*）、橙翅噪鹛（*Trochalopteron elliotii*）、灰胸薮鹛（*Liocichla omeiensis*）、四川旋木雀（*Certhia tianquanensis*）、滇䴓（*Sitta yunnanensis*）、白脸䴓（*S. przewalskii*）和蓝鹀（*Emberiza siemsseni*）。

西南山地（横断山区）鸟类研究简史

从19世纪中叶第一次鸦片战争开始，以外国传教士为代表的西方研究者开始逐渐进入这片宝地，针对该区域鸟类的考察和采集工作也由此展开。其中最具代表性的人物首推法国神父谭卫道（Armand David），他于1869年2月28日抵达四川穆坪（今宝兴），至当年11月22日离开，在前后近9个月的时间内采获了大量的动植物标本，计有676种植物、441种鸟类和145种哺乳动物，其中就包括大熊猫（*Ailuropoda melanoleuca*）、绿尾虹雉（*Lophophorus lhuysii*）、珙桐（*Davidia involucrata*）等本区域所特有的珍贵物种。谭卫道作为首位进入到青藏高原东南缘进行系统采集的博物学家，取得了丰硕的成果，他本人也因这些重要发现被法兰西地理学会、学者联盟等学术团体联合授予了一枚金质奖章。

谭卫道的工作首次向世界揭示了我国西南山地独特而又丰富的生物多样性，也激励了许多慕名而来的各国人士。英国鸟类学家莱昂内尔·沃尔特·罗斯柴尔德勋爵（Lionel Walter Rothschild）、英国植物学家乔治·福雷斯特（George Forrest）和美籍奥地利探险家约瑟夫·洛克（Joseph Rock）为其中的代表人物。罗斯柴尔德勋爵于1926年发表了《云南鸟类区系》（*The Avifauna of Yunnan*），是对此前有关云南鸟类研究最为全面而系统的整理。福雷斯特从1904年起，直至1932年逝于云南，近28年间先后七次在云南西部及周边进行采集。除了主要的植物标本收集之外，他还获取了上万号的鸟类标本，今天的四川柳莺（*Phylloscopus forresti*）最早就是由罗斯柴尔德勋爵于1921年依据福雷斯特在丽江采集到的1号雌鸟标本进行描述的。洛克于1922年由缅甸进入云南，一路向北进入丽江，并于1924年最终到了四川木里。洛克是唯一一位在中国西南边境地区长期生活及考察、一生大部分时间都生活在云南的西方人。在华期间除了近6万号植物标本之外，他还收集到了1600号鸟类标本，对丽江纳西族的文化和历史也颇有研究。

除上述研究者之外，美国人珍·鲍尔德斯顿女士（Jane C. Balderston）于1919至1949年间长期生活在四川成都，她以华西协合大学校园内住所为中心，在方圆十余公里的区域内坚持观察记录鸟类。1938年夏和1943年7月，她还先后利用到邛崃山脉中的峨眉山和蒙顶山避暑之际，对当地的鸟类进行了仔细观察，并于1946年在《华西边疆研究学会杂志》（*Journal of the West China Border Research Society*）上正式发表了自己的记录。参照《大英百科全书》上对于观鸟的定义——"在自然环境中对野生鸟类进行观察的一种兼具科学性的流行休闲方式"，珍女士所著此文应是在这一区域进行现代意义上的观鸟活动并详加记载的开山之作。

而早年间的中国本土鸟类学先贤们，尽管面临时局动荡、财力物力有限等窘境，依然克服了令人难以想象的困难，开展了一些研究工作。1922年8月18日，中国科学社生物研究所在南京市成贤街文德里社址宣告成立，这是中国近代史上第一个由本国学者创办的生物学研究机构。1929年4月，时任生物研究所所长的秉志先生[1]即委派徐锡藩和刘子刚入川采集动植物标本，前后历时近6个月，足迹涉及峨眉、峨边、雅安和康定等地，由此也标志着我国学者在四川乃至西南山地独立进行生物学研究的开始。其时供职于静生生物调查所的寿振黄先生[2]对这批鸟类标本进行了分类整理，先后在《静生生物调查所汇报》（*Bulletin of the Fan Memorial Institute of Biology*）上发表了两篇论文，这也是中国鸟类学家关于横断山区（西南山地）鸟类最早的研究报告。1930年下半年卢作孚先生[3]在重庆北碚创建了中国西部科学院，是近代以来西部地区建立最早的综合性科学研究机构，也是当时国内为数不多的私立科研单位之一（潘洵，2005）[4]。西部科学院成立后，于1933至1934年间先后在四川西部、南部和中部进行了采集工作，所获鸟类标本由王希成先生整理，以《四川鸣禽类之研究》为题在《中国科学社生物研究所丛刊·动物学系列》（*Contribution from the Biological Laboratory of*

---

[1] 秉志（1886—1965），字农山，河南开封人，我国现代动物学的开创者和主要奠基人。
[2] 寿振黄（1899—1964），字理初，浙江诸暨人，我国脊椎动物学研究的开创者，鸟类学研究的奠基人之一。
[3] 卢作孚（1893—1952），四川省重庆府合州县（今重庆市合川区）人，著名爱国实业家、教育家。
[4] 潘洵：《中国西部科学院创建的缘起与经过》，载于《中国科技史杂志》，2005年第1期，第19—26页。

摄影：董磊

the Science Society of China, Zoological Series）上发表。

新中国成立之后，国家先后组织了一些针对横断山区（西南山地）的大型综合考察。如1959至1961年间，在四川南部、西部和云南西北部进行的"南水北调综合考察"，中国科学院动物研究所、昆明动物研究所、南充师范学院（今西华师范大学）等单位均派员参与。1981至1985年，中国科学院青藏高原综合科学考察队进入西南山地，对云南西部、北部，四川西部、南部和西藏昌都地区东南部进行了系统调查，先后有唐蟾珠、魏天昊、郑宝赉、杨岚、徐延恭等参加，考察所涉及范围近20万km²，根据所获取的大量一手资料，最终由中国科学院动物研究所唐蟾珠、徐延恭和昆明动物研究所杨岚三位先生整理撰写成为《横断山区鸟类》一书并于1996年出版。考察队员们白天在高山密林中采集标本、收集数据，时常还要在昏暗的烛光下制作标本至深夜，条件艰苦之外，有时甚至还会有危及生命的风险。来自昆明动物研究所的著名动物学家彭鸿绶先生在参与考察过程中，于1981年8月24日在云南中甸（今香格里拉）突发高山急性肺水肿，不幸以身殉职。

除上述大型考察之外，四川农业大学李桂垣先生及其同事自1959年开始坚持在四川西部进行鸟类调查并对区系组成及演替进行了较为深入的研究。昆明动物研究所则组织人员于1973至1975年间对高黎贡山地区进行了专门的调查。以上跨越百年的由先辈们历经艰辛、排除万难所完成的工作奠定了我们今天对于横断山区（西南山地）鸟类多样性研究与保护的宝贵基础。

鸟类结构图示

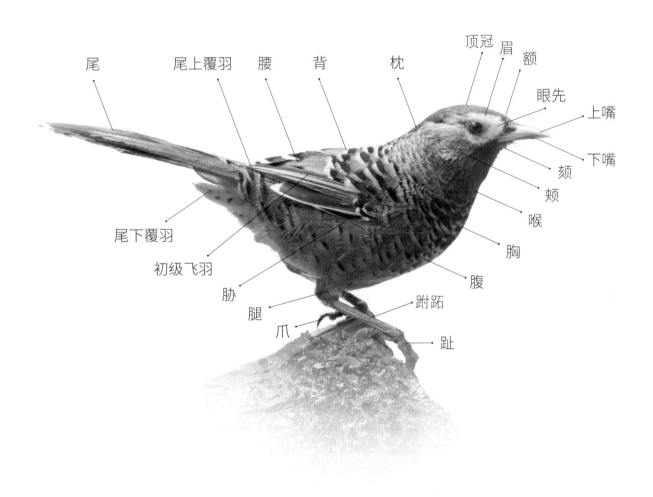

斑背噪鹛　摄影：何屹

鸟类

Birds

1. 青头潜鸭　Baer's Pochard

雁形目　Anseriformes　**鸭科**　Anatidae　*Aythya baeri*

体长约46cm的中型鸭科鸟类。雌雄两性于野外区别较大。雄鸟头部青绿色，白色虹膜于头部甚为明显；颈部棕褐色，与青绿色喉部形成鲜明对比，两胁带有灰白色及褐色斑纹；背部深棕色；尾下覆羽为白色。雌鸟似雄鸟，但整体羽色较为暗淡，头部颜色为深褐色，虹膜近黑色。

100多年前，青头潜鸭曾广泛分布于中国、俄罗斯、印度、缅甸、泰国等地。20世纪50年代，中国境内青头潜鸭的记录仅在东北、山东、长江中下游和福建出现。80年代后期，中国西南的西藏、四川、云南有零星记录。2011年以来，中国西南的记录仅有四川和云南，其中四川的川西平原的湖泊和河流中，尤其是城市湿地环境，有多笔个体越冬记录，包括成都市、广汉市、德阳市、绵阳市、以及隆昌县等。近年来，青头潜鸭野外种群数量急剧下降，已被列为极危物种，全球种群数量推测低于1000只。青头潜鸭常于多芦苇的淡水湖泊和沼泽环境繁殖，冬季栖息于淡水湖泊、水塘和水库等水域，有时与其他潜鸭类混群活动；以水草为食，有时亦捕捉小型水生昆虫、蛙类等。

雌鸟　摄影：何既白

Baer's Pochard is a medium-sized duck of the Anatidae family, with a body length of about 46cm, sexually dimorphic. The male bird has bluish green head with conspicuous white iris on the head and the bluish green throat in sharp contrast to the brown neck. The two flanks are stained with grey and brown spots, the back is dark brown and the undertail coverts are white. The female bird resembles the male, but the general color of plumage is relatively darker. The head is dark brown and the iris is almost black.

Baer's Pochard was once widely distributed in China, Russia, India, Myanmar, Thailand and so on over 100 years ago. In the 1950s, records of the Baer's Pochard in China were only made in the northeast China, Shandong, the middle and lower reaches of the Yangtze River, and Fujian. In the late 1980s, there were sporadic records in Xizang, Sichuan, and Yunnan in southwest China. Since 2011, records in Southwest China have only been made in Sichuan and Yunnan, among which several records of individuals overwintering in lakes and rivers in the Western Sichuan Plain, especially in the wetland environment in cities, including Chengdu, Guanghan, Deyang, Mianyang, and Longchang County, etc., were documented. In recent years, the wild population of the Baer's Pochard has dropped drastically and has been classified as a critically endangered species. Its global population is estimated to be less than 1,000. The Baer's Pochard often breeds in freshwater lakes and marshes with dense reeds bush and inhabits in freshwater lakes, ponds, and reservoirs in winter, and sometimes mixes with other pochards. It feeds on aquatic plants and sometimes small aquatic insects and frogs.

左雌鸟右雄鸟　　摄影：董磊

2. 中华秋沙鸭 Scaly-sided Merganser

雁形目 Anseriformes 鸭科 Anatidae *Mergus squamatus*

体长约57cm的大型鸭科鸟类。雌雄两性于野外区别较大。雄鸟头部、颈部墨绿色而具金属光泽，头顶有明显羽冠，狭长的喙部为鲜亮的橙红色，喙尖带钩，胸部白色几无斑纹，两胁白色具黑色鱼鳞状斑纹；背部多墨绿色，两翼具显著的白色翼镜。雌鸟羽冠较短，头颈部褐色，背部和胸部灰色具鳞片状斑纹。亚成年的雄性可通过更加暗色的翼上覆羽与雌性区别。

中华秋沙鸭主要繁殖于俄罗斯西伯利亚东南部和中国东北地区，中国也是中华秋沙鸭种群世界范围内最重要的越冬地，四川都江堰冬天也较为稳定地有中华秋沙鸭种群出现，冬季还记录于日本、韩国、泰国、缅甸、越南。常于针叶林、针阔混交林周边的河流两岸繁殖，冬季栖息于河流、水库等环境。主要以鱼类为食，也采食水生昆虫的幼虫、虾蟹等，还有记录展示了中华秋沙鸭团队协作分工捕鱼的场面，有的冲浪，有的潜水，将鱼群赶到浅滩。越冬季节的取食高峰多在上午8点至9点，以及10点至11点，还有午后的14点到15点，其中10点到11点为最活跃的时段，下午多在休息。环境温度高于10度，游泳行为更多；低于10度，取食行为更常见。常成小群或零星个体活动，少有10只以上集群活动，性格机警，稍有惊动则起飞，多沿河流往上游飞走。河流两岸和河中露出的石滩是中华秋沙鸭重要的栖息环境，并且需要避开人类干扰，其夜间栖息环境为河边草地、灌丛或柳树下。通常很安静，雌性有时发出"亏，克"声。

Scaly-sided Merganser is a big bird of the Anatidae family, with a body length of about 57cm, sexually dimorphic. The male bird has dark green head and neck, where the feathers show metallic luster. It has a prominent crest and a bright reddish orange bill which is narrow and long, with a hook on the tip of the bill. The breast is white with almost no markings and both flanks are white with black squamates that resemble fish scales. The back is mainly dark green, with prominent white speculums on both wings. The female has a shorter crest, brown head and neck, grey back and breast with squamates. The sub-adult males can be distinguished from females by their darker wing coverts.

Scaly-sided Merganser mainly breeds in Southeastern Siberia of Russia and Northeast China. China is also the most important wintering ground for the Scaly-sided Merganser population worldwide. Its population appears stably in Dujiangyan, Sichuan in winter. There are also overwintering records made in Japan, South Korea, Thailand, Myanmar, and Vietnam. It often breeds on banks of rivers around coniferous forests and coniferous and broad-leaved mixed forests, and inhabits rivers and reservoirs in winter. It mainly feeds on fish, and also on the larvae of aquatic insects, shrimps, and crabs. Scenes of Scaly-sided Merganser collaboratively working at fishing were once spotted—some surfing, and others diving and driving the fish to the shallows. In winter, feeding usually peaks from 8 a.m. to 9 a.m. and 10 a.m. to 11 a.m., and 2 p.m. to 3 p.m.. During these peak windows, 10 a.m. to 11 a.m. is when the Scaly-sided Mergansers are most active and they usually rest in the afternoon. It shows more swimming behavior if the temperature is higher than 10 degrees Celsius and more feeding behavior below 10 degrees Celsius. It often lives in small groups or by sporadic individuals. Group members are rarely more than 10. They are vigilant and tend to respond to the slightest threat by taking off and mostly fly to upstream along the river. The gravel beaches on river banks or exposed in the river are important habitats for the Scaly-sided Merganser and human interference needs to be avoided in such areas. Its night habitats include grasslands, shrubs, and areas under willow trees by the river. It is usually quiet, and females sometimes make a call of qua, ke.

左雌鸟右雄鸟　　摄影：汤开成

3. 斑尾榛鸡　Chinese Grouse

鸡形目　Galliformes　雉科　Phasianidae　*Tetrastes sewerzowi*

体长约35cm的中型雉科鸟类。雌雄两性于野外容易区别。雄鸟上体棕褐色，具黑色横斑；颏部、喉部黑色，具白色边缘；胸腹部及两胁棕色，密布黑色横斑。雌鸟似雄鸟，但体色较暗；颏部、喉部褐色且具白色细纹。

斑尾榛鸡为更新世冰期耐寒动物群欧亚大陆雉鸡类的后代，适应严酷的气候和特殊的冰雪环境，现为青藏高原东缘2400—4300米的高山针叶林特有鸟种，见于甘肃、青海东北部、四川北部和西部、云南西部和西藏东部。成年斑尾榛鸡为较为严格的植食性鸟类，主要以柳和桦的芽叶以及云杉种子为食，辅以其他植物的花、叶、嫩枝梢，嘴的形态已进化得适应于可以一次性啄取整个柳芽，而幼鸟则多取食蚂蚁卵、甲虫，以及植物嫩芽。

雌鸟　摄影：唐军

Chinese Grouse is a medium-sized bird of the Phasianidae family with a body length of about 35cm. It is easy to distinguish between male and female in the wild. The upperparts of male are brown with black transverse spots; chin and throat are black with white edges; the breast, belly, and flanks are brown with dense black transverse spots. The female resembles the male bird, but the general color is duller, the chin and throat are brown with white fine lines.

Chinese Grouse is the descendant of grouses in Eurasia continent of the Pleistocene (ice-age) fauna, adapted to harsh climates and special snow and ice environments. It is now a unique bird species of the alpine coniferous forests at the eastern edge of the Qinghai-Xizang Plateau from 2400m to 4300m. It is found in Gansu, northeastern Qinghai, northern and western Sichuan, western Yunnan, and eastern Xizang. The adult Chinese Grouse is strictly herbivorous. It mainly feeds on the spruce seeds, buds and leaves of willow and birch, supplemented by flowers, leaves, and twigs of other plants. The morphological evolution of its bill is adapted to be able to peck the whole willow bud at once, while young birds mostly feed on ant eggs, beetles, and plant shoots.

雄鸟　摄影：董磊

4. 红喉雉鹑 Chestnut-throated Monal Partridge

鸡形目　Galliformes　雉科　Phasianidae　*Tetraophasis obscurus*

体长约50cm的中型雉科鸟类。雌雄两性于野外较难区别。成鸟头灰色，眼周具非常明显的红色裸皮；喉部栗红色，外缘常呈白色或皮黄色；胸部灰色具黑色点状斑纹，腹部灰褐色具棕黄色和白色斑纹；背部及两翼为灰褐色具白色斑纹。

红喉雉鹑为中国特有留鸟，见于四川省岷山—邛崃山系、青海南部、甘肃祁连山等地，较黄喉雉鹑分布范围更为狭窄，常栖于高海拔的高山针叶林和林线之上的杜鹃灌丛地带，喜食贝母、青稞等植物以及昆虫，有时发出连续的"维欧-维欧"的叫声，具颤音。

Chestnut-throated Monal Partridge is a medium-sized Phasianidae bird of about 50cm in length. It is difficult to distinguish between male and female in the wild. Adult birds have grey heads with very obvious red bare skin around the eyes, the chestnut-red outer edges of the throat often have white or buffy skin, the breast is grey with black spots, the belly is greyish brown with brownish yellow and white markings, the back and wings are greyish brown with white markings.

Chestnut-throated Monal Partridge is a resident bird endemic to China. It is found in the Min-Qionglai mountain system of Sichuan as well as southern Qinghai and Qilian Mountains of Gansu. It has a narrower distribution than that of Buff-throated Monal Partridge. It often inhabits high-altitude mountain coniferous forests and Rhododendron shrubs above the tree line. This bird feeds on Fritillaria, highland barley and other plants and insects. Sometimes it makes a continuous rolling vio-vio call.

摄影：罗平钊

5. 黄喉雉鹑 Buff-throated Monal Partridge

鸡形目　Galliformes　雉科　Phasianidae　*Tetraophasis szechenyii*

　　体长约48cm的中型雉科鸟类。雌雄两性于野外较难区别。成鸟头灰色，眼周具十分明显的红色裸皮；喉部棕黄色；胸部深灰色具黑色点状斑纹，腹部灰色具棕黄色斑纹；背部及两翼灰褐色沾棕色具白色斑纹。

　　黄喉雉鹑为中国特有鸟种，见于四川西部的甘孜州、青海玉树以南、云南西北部、西藏的昌都和林芝等地区；常栖于高海拔的针叶林、高山灌丛和林线上的岩石苔原地带，尤其喜好高山栎林，冬季下至中高海拔地区的混交林和林缘地带活动。它们喜食贝母、青稞等植物。

　　Buff-throated Monal Partridge is a medium-sized Phasianidae bird of about 48cm in length. It is difficult to distinguish between male and female in the wild. Adult birds have grey heads with very obvious red bare skin around the eyes; the throat is brownish yellow; the chest is dark grey with black markings; the abdomen is grey with brownish yellow markings; the back and wings are greyish brown with rufous tinges and white markings.

　　Buff-throated Monal Partridge is a bird endemic to China, found in Ganzi Prefecture in western Sichuan, south of Yushu in Qinghai, northwestern Yunnan, Changdu and Linzhi in Xizang. It often inhabits high-altitude coniferous forests, alpine shrubs, and rocky tundra areas above the tree line, especially prefers alpine oak forests. In winter, it moves down to mixed forests and forest edges in middle and high-altitude areas. It likes to eat Fritillaria, highland barley, and other plants.

摄影·董磊

6. 四川山鹧鸪 Sichuan Partridge

鸡形目　Galliformes　雉科　Phasianidae　*Arborophila rufipectus*

体长约30cm的小型雉科鸟类。雌雄两性羽色区别较大。雄鸟前额白色，头顶至枕部红褐色多黑色斑纹，红色眼圈外眼周黑色，脸颊红褐色环绕黑斑；喉部白色少斑纹，胸部栗红色，腹部白色；背部多灰色斑纹，两翼具棕红色、灰色斑纹。雌鸟似雄鸟，但头顶、脸颊灰褐色，胸部深褐色。

四川山鹧鸪是中国西南山地特有珍稀留鸟，自然分布地域极其狭窄，主要见于四川省甘洛、峨边、马边、沐川、屏山、雷波等小相岭至大凉山区域，及云南省绥江、永善等县的海拔1000—2300米的山地森林，尤其是亚热带天然阔叶林或针阔混交林，也选择人工林，不喜好茂密的次生灌丛及人工幼林生境，冬季迁至低海拔区域越冬，选择平缓坡、草本植物低矮稀疏、乔木稀疏、距小路和林缘较近的栖息地，并倾向于选择距水源较近且地表落叶丰富的环境，主要以植物果实、种子为食。

雌鸟　摄影：李祺

雄鸟　摄影：戴波

Sichuan Partridge is a small bird of the Phasianidae family with a body length of about 30cm. The male and female are quite different in plumage. The male bird has a white forehead and from the cap to the nape is reddish brown with many black markings, the area around the red eye ring is black and the cheeks are reddish brown with dark spots, the throat is white and has little markings, the breast is chestnut and the abdomen is white, the back has many grey markings, and there are brownish red and grey wing bars on the wings. The female bird resembles the male, but the cap, cheeks, and breast are darkish brown.

The natural distribution area of this species is extremely narrow. It is a rare resident bird endemic to the mountainous regions of southwest China, mainly found in mountain forests with an altitude of 1000m to 2300m in Ganluo, Ebian, Mabian, Muchuan, Pingshan, Leibo, and other areas from Xiaoxiangling to Daliangshan in Sichuan, and Suijiang, Yongshan of Yunnan, especially in subtropical natural broad-leaved forests or coniferous and broad-leaved mixed forests, but it also chooses man-made forests. This bird does not like dense secondary shrubs and man-made young forest habitats. In winter it moves down to low-altitude areas, choosing habitats with gentle slopes, low and sparse herbs, sparse trees, which are close to trails and forest edges, and tending to choose environments close to water sources and rich in deciduous leaves on the ground. It mainly feeds on fruits and seeds.

7. 灰胸竹鸡 Chinese Bamboo Partridge

鸡形目　Galliformes　雉科　Phasianidae　*Bambusicola thoracicus*

　　体长约32cm的中型雉科鸟类。雌雄两性于野外较难区别，仅雌性个体略小。成鸟额头蓝灰色，头顶至枕部褐色，眉纹蓝灰色后延至颈部，脸颊棕褐色；喉部棕色，胸部蓝灰色，腹部棕黄色，两胁具显著的心形黑色点状斑纹；背部、两翼灰褐色具棕褐色斑纹。

　　灰胸竹鸡为中国特有留鸟，见于陕西省秦岭以南，四川盆地以东、中国长江以南的中低海拔地区，常栖于山地丘陵和临近的平原地带，昼出夜伏，善于跳跃和奔跑，飞行能力弱，见于各种林地环境，尤以竹林、灌丛和草丛地带为多。灰胸竹鸡属杂食动物，采食量少但频繁，以各种植物种子、嫩芽、果实、谷粒和其他农作物以及昆虫、蠕虫等为食。

Chinese Bamboo Partridge is a medium-sized bird of the Phasianidae family with body length of about 32cm. It is difficult to distinguish between male and female in the wild, with the only difference that females are slightly smaller. Adult birds have a bluish grey forehead, brown from the cap to the nape, bluish grey eyebrows extending to the neck, and the cheeks are rufous. The throat is brown, the chest is bluish grey, and the abdomen is brown, both flanks have prominent heart-shaped black markings, the back and two wings are greyish brown with rufous markings.

Chinese Bamboo Partridge is a resident bird endemic to China, found in low and middle altitudes in south of the Qinling Mountains of Shaanxi Province and south of the Yangtze River to the east of the Sichuan Basin. They usually inhabit mountainous hills and adjacent plains. They are active in the daytime and roost at night, very good at hopping and running, but with poor ability of flight, and found in various woodland habitats, especially bamboo thickets, shrubs, and grasslands. Chinese Bamboo Partridge is omnivorous with low but frequent feed intake. It feeds on various seeds, shoots, fruits, grains, and other crops as well as insects and worms.

摄影：唐军

8. 红腹角雉 Temminck's Tragopan

鸡形目　Galliformes　雉科　Phasianidae　*Tragopan temminckii*

　　雄鸟体长约65cm，雌鸟体长约55cm的中型雉科鸟类。雌雄两性羽色区别明显。雄鸟羽色艳丽，前额至头顶黑色，眼周具一亮蓝色裸皮，脸颊黑色，喙部深灰色；下颌深蓝色，喉部、颈侧黑色，颈部黄褐色；胸部橙红色，腹部、背部、两翼红色具灰白色圆形斑点；尾羽深褐色。雌鸟整体暗淡偏灰褐色；胸腹部、背部多浅色斑点。

　　红腹角雉在中国见于西藏、云南、四川、贵州、甘肃、陕西等地区，国外分布于印度东北部、缅甸东北部和越南极西北部，为各地留鸟。有研究表明，红腹角雉偏向于中高海拔的山地常绿落叶阔叶混交林，不太喜欢干燥茂密的竹林生境。最常见啄食植物嫩叶、幼芽、花、茎、果实和种子，秋季有发现取食竹笋，也刨食落叶层及松土层内的落果、种子、无脊椎动物及植物的根，有时还追捕节肢动物，食性随地区和季节不同而有所变化，未见其对某些食物有所依赖，推测是该物种较该属其他种数量更大、分布更广和适应力更强的原因之一。另外，据观察，山地阴坡分布着更多喜食的动植物，加之阴暗潮湿的环境，故红腹角雉偏爱活动于阴坡，并有随季节变化垂直迁移的习性，多会避开积雪的区域。夏秋季，雄性会发出非常有代表性的"哇哇"叫声，似小孩的啼哭，故俗称"娃娃鸡"。

　　Temminck's Tragopan is a medium-sized bird of the Phasianidae family with a body length of about 65cm for male birds and 55cm for female birds. The male and female plumage are distinct. The male bird has a bright plumage, black forehead cap, bright blue bare skin around the eyes, black cheek and dark grey bill; dark blue chin, black throat and side of neck, and yellowish brown neck, orange-red chest; the abdomen, back, and wings are red with greyish white round spots; the tail feathers are dark brown. The general plumage of female bird is dark and greyish brown, there are many pale-colored spots on the chest, the abdomen and the back.

　　Temminc's Tragopan is found in Xizang, Yunnan, Sichuan, Guizhou, Gansu, Shaanxi, and other regions in China. It also distributes abroad in northeastern India, northeastern Myanmar, and northwestern Vietnam, and a resident bird to all these regions. Studies have shown that Temminck's Tragopan prefers mixed evergreen and deciduous broad-leaved forests in medium and high altitude mountains, and does not like the dry and dense bamboo thickets habitat. Commonly eating young leaves, buds, flowers, stems, fruits and seeds of plants, it was also discovered eating bamboo shoots in autumn, as well as fruit, seeds, invertebrates and plant roots in the deciduous and loose soil layer. Sometimes arthropods are also being hunted down.

Their feeding habits vary within different regions and seasons, and they are not dependent on certain foods. It is speculated that this is one of the reasons for its larger number, wider distribution, and stronger adaptability than other species of this genus. In addition, it has been observed that there are more food choices which the bird prefers by the shady slopes of the mountains, as well as the dark and humid environment, the Temminck's Tragopan prefers to move on the shady slopes. It also has the habit of seasonal vertical migration, and often avoids snowy areas. In summer and autumn, male bird will make a very representative wow-wow cry, like the cry of a child, so it is commonly called "Wawaji (means baby chicken)".

雄鸟　摄影：唐军

雌鸟　摄影：钟宏英

9. 白尾梢虹雉 Sclater's Monal

鸡形目 Galliformes 雉科 Phasianidae *Lophophorus sclateri*

体长约65cm的大型高山雉科鸟类。雌雄两性羽色区别较大。雄鸟羽色艳丽，周身具闪耀的金属光泽，头顶为金属绿色，眼周具一蓝色裸皮，脸颊深绿色，喙部象牙白色；颈侧、枕部亮棕色；颈部、胸部、腹部及尾下覆羽深绿色；背部及尾上覆羽为白色；尾羽具大面积棕色，尾羽末端为明显的白色。雌鸟整体暗淡、偏棕褐色；眼周具一蓝色裸皮，下颏白色；尾羽末端为较为明显的白色。

白尾梢虹雉野外只生活在喜马拉雅山脉东部和西南山地中西部，具体包括缅甸，印度东北部，中国西藏东南部的米林市西南部、易贡藏布下游、丹巴曲上游山脉和伯舒拉岭，云南西北的高黎贡山和怒山山脉（原碧罗雪山）。常栖于海拔2500m以上的高山森林和亚高山针叶林、林缘灌丛、草甸等地带。主要以草籽及一些植物的果实、种子为食，根据高黎贡山的粪便分析，春季白尾梢虹雉至少采食23种植物，不同月份植物种类有所差异，其中苔藓和尼泊尔绿绒蒿所占比例相对较高。

Sclater's Monal is a large alpine bird of the Phasianidae family with a body length of about 65cm. Male and female plumage are quite different. Male bird feathers are gorgeous, with shiny metallic luster around the body, metallic green on the cap, blue bare skin around the eyes, dark green cheeks, ivory white bill, bright brown on sides of neck and nape. The neck, chest, abdomen, and under tail coverts are dark green, the back and upper tail coverts are white, the tail feathers have a large area of brown, and the tips are significantly white. The general plumage of female bird is dull rufous. There is a blue bare skin around the eyes, the chin is white, the tips of tail feathers are significantly white.

Sclater's Monal only lives in the eastern Himalayas and the central and western parts of the Mountains of Southwest China, as well as Myanmar, northeastern part of India, southwestern part of Milin city, southeastern part of Xizang, China, lower reach of Yigong Zangbu, upper Danbaqu mountain range, and the Bashula Mountain, Gaoligong Mountain, and Nushan Mountain (formerly Biluo Snow Mountain) in northwestern Yunnan. They often live in alpine forests and sub-alpine coniferous forests, forest edge shrubs, meadows, and other areas which is around 2500m. It mainly feeds on grass seeds and fruits and seeds of some plants. According to the analysis of manure in Gaoligong Mountain, the Sclater's Monal eats at least 23 species of plants in spring, and the species varies in different months, the moss and *Meconopsis napaulensis* have a high proportion.

雄鸟　　摄影：王斌

雌鸟　　摄影：董磊

10. 绿尾虹雉 Chinese Monal

鸡形目　Galliformes　雉科　Phasianidae　*Lophophorus lhuysii*

雌鸟　摄影：董磊

体长约78cm的大型雉科鸟类。雌雄两性羽色区别较大，雄鸟羽色艳丽，周身闪耀金属光泽，头部为金属绿色，眼先具一蓝色裸皮；颈侧、枕部亮棕色；胸部、腹部、尾下覆羽深绿色；背部白色，但多被蓝紫色两翼遮挡；尾羽金属绿色。雌鸟整体暗淡，偏棕褐色；下颌白色；腰部白色。

绿尾虹雉为西南山地特有留鸟，主要分布在四川西部岷山、邛崃山、相岭、大雪山和沙鲁里山等山系，并边缘性见于云南西北部、西藏东部、甘肃南部和青海东南部，曾在四川唐家河自然保护区海拔约1300米的河谷记录过一只。甘肃白水江自然保护区的研究结果表明，绿尾虹雉多栖息在海拔2800米以上的高山草甸、悬崖峭壁、针叶林等地带，其自然条件恶劣，坡度极为陡峭，局部地方超过80度。一般阳坡较平缓，为高山草甸；阴坡较陡，为针叶林。

Chinese Monal is a large bird of the Phasianidae family with a body length of about 78cm. There is a big difference in plumage between male and female birds. Male birds have beautiful and shiny plumage with metallic luster, the head is metallic green and the lores have a blue bare skin; the neck and nape are bright brown; the breast, abdomen, and undertail coverts are dark green; the back is white, but mostly blocked by blue and purple wings; the tail feathers are metallic green. The general plumage of female bird is dark and brown, with white lower chin and waist.

Chinese Monal is a resident bird endemic to the Mountains of Southwest China mainly distributed in the mountain systems of Min, Qionglai, Xiangling, Daxueshan, and Shaluli in western Sichuan, and marginally seen in northwestern Yunnan, eastern Xizang, southern Gansu and southeastern Qinghai, with one record in a valley of about 1300m in Tangjiahe Nature Reserve, Sichuan. According to the research results of the Baishuijiang Nature Reserve in Gansu, the Chinese Monal inhabits alpine meadows, cliffs, and coniferous forests above 2800m. The natural conditions are harsh, and the slopes are mostly steep, locally exceeding 80 degrees. Generally, the sunny slope is relatively gentle, which is alpine meadow, and the shade slope is relatively steep, which is coniferous forest.

雄鸟　摄影：董磊

11. 白马鸡 White Eared Pheasant

鸡形目　Galliformes　雉科　Phasianidae　*Crossoptilon crossoptilon*

体长约83cm的大型雉科鸟类。雌雄两性于野外较难区别。成鸟整体近白色；头顶黑色，眼周具红色裸皮，耳羽簇白色；喉部、颈部、枕部、背部、胸部、腹部皆为白色，飞羽黑色，尾羽黑色且呈蓬松丝状。

白马鸡是中国西南山地特有留鸟，见于四川西部、云南西北部、青海东南部和西藏东部高山地区。栖息于多种生境类型中，包括针阔混交林、高山针叶林、高山灌丛草甸以及镶嵌在各垂直植被中的高山栎林，尤好在开阔的林间空地和林缘地带活动。

白马鸡是杂食性鸟类，主要取食植物，春季取食绿色植物的嫩芽和叶，夏季选择茎、叶，秋季偏好花蕾、果实和草籽，冬季则吃枯叶和植物根部。亦食昆虫幼虫、蠕虫。秋末和冬季结群到农田中采食马铃薯、荞麦、旱芹、油菜、青稞等。

White Eared Pheasant is a large bird of the Phasianidae family with a body length of about 83cm. It is difficult to distinguish between male and female in the wild. The adult bird is almost white in general; the cap is black, the bare skins around eyes are red, the ear coverts are white; the throat, neck, nape, mantle, breast and abdomen are all white; the flight feathers are black and the tail feathers are black, fluffy, and silky.

White Eared Pheasant is a resident bird endemic to the Mountains of Southwest China, found in western Sichuan, northwestern Yunnan, southeastern Qinghai and alpine areas of eastern Xizang. It inhabits in a variety of habitat types, including coniferous and broad-leaved mixed forests, alpine coniferous forests, alpine shrubs meadows and alpine oak forests embedded in various vertical vegetations, especially in open forest clearings and forest edges.

White Eared Pheasant is omnivorous, mainly feeding on plants. In spring it feeds on the buds and leaves of green plants; in summer stems and leaves are prefered. It eats flower buds, fruits, and grass seeds in autumn, and dead leaves and plant roots in winter. It also feeds on insect larvae and worms. At the end of autumn and winter, they gather in the farmland to eat potatoes, buckwheat, celery, brassica, hulless barley, etc..

摄影：董磊

12. 蓝马鸡 Blue Eared Pheasant

鸡形目　Galliformes　雉科　Phasianidae　*Crossoptilon auritum*

　　体长约85cm的大型雉科鸟类。雌雄两性于野外较难区别。成鸟整体蓝灰色；头顶深褐色，眼周具大面积红色裸皮，耳羽簇白色向下延伸；颈部、枕部、胸部、腹部、背部、尾羽皆为蓝灰色。

　　蓝马鸡是中国特有留鸟，仅见于青海东部和东北部，甘肃西北部及南部，四川西北部，以及与其他栖息地相隔甚远的宁夏北部贺兰山，为典型的亚高山针叶林和高山灌丛鸟类，栖息地包括云杉林、山杨桦木林、杜鹃灌丛、针阔混交林、高山针叶林和高山草甸。

　　蓝马鸡最喜采食植物的嫩茎和叶，其次是花和块根。在秋冬季主要取食种子、块根和凋落的枯叶等，繁殖期也啄食昆虫为身体提供额外的营养。在蓝马鸡觅食区域，常能看到刨开的小土坑。

Blue Eared Pheasant is a large bird of the Phasianidae family with a body length of about 85cm. It is difficult to distinguish between male and female in the wild. Adult birds have general plumage of bluish grey, the cap is dark brown, the bare skin around eyes is red, and the white ear feather tufts extend downward; the neck, nape, breast, abdomen, back and tail feathers are all bluish grey.

Blue Eared Pheasant is a resident bird endemic to China. It is only found in eastern and northeastern Qinghai, northwest and south of Gansu, northwest of Sichuan, and Helan Mountain in northern Ningxia which is far away from other habitats. It is a typical bird in sub-alpine coniferous forests and alpine shrubs. Its habitats include spruce forests, aspen birch forests, rhododendron shrubs, coniferous and broad-leaved mixed forests, alpine coniferous forests, and alpine meadows.

Blue Eared Pheasant prefers to feed on the tender stems and leaves of plants, followed by flowers and tubers. In autumn and winter, the birds mainly feed on seeds, tubers and littered leaves, while during the breeding season, insects are also taken to provide additional nutrients. In the foraging ground of Blue Eared Pheasant, small gouged holes can be seen often.

摄影：罗平钊

13. 红腹锦鸡 Golden Pheasant

鸡形目 Galliformes 雉科 Phasianidae *Chrysolophus pictus*

体型修长而矫健的中型雉鸡，体长约1米，其中尾可接近40厘米。雌雄两性羽色差异甚大，雄鸟是雉科家族中羽色最为艳丽的成员。雄鸟的羽色华丽，头顶金黄色丝状冠羽，上背部暗绿色，其余上体金黄色，颈部羽毛呈橘黄色，羽缘黑色，形成虎纹般的披肩。下体鲜红色，尾羽黄褐色，布满黑色斑点。雌性羽色以灰褐色和黄褐色为主，布满长短不一的横斑纹，脚黄色。

红腹锦鸡主要栖息于海拔3000米以下的疏林灌丛、针阔混交林和阔叶林地带，偏好多岩石的山坡，森林郁闭度①高，林下空旷，有树丛或竹丛的环境，甚至农田环境，对山坡的朝向没有明显的偏好。仅见于中国，分布于青海、甘肃、宁夏、陕西、河南、四川、西藏、湖北、湖南、云南、贵州和广西，在四川邛崃山系发现有与白腹锦鸡自然杂交的后代。

① 郁闭度：林分整个林木林冠的投影面积与此林地总面积的比值，用十分法表示，常作为控制抚育采伐强度的指标，也是区分有林地、疏林地、未成林造林地的主要指标。林分是指内部特征基本一致而与周围又有显著差异的森林地段。

雌鸟　摄影：路家兴

雄鸟　摄影：叶昌云

Golden Pheasant is a slender and vigorous medium-sized pheasant with a body length of about 1m of which the tail can approach 40cm. There is a big difference in plumage between male and female birds, and male birds are the most colorful members of the Phasianidae family. The plumage of male is gorgeous, with golden yellow silky crest on the cap, dark green mantle, and the rest of the upperpart is golden yellow. The neck coverts are orange with black edges, forming a tiger-striped cape. The underpart is bright red, and the tail feathers are yellowish brown covered with black spots. The plumage of female is mainly greyish brown and yellowish brown, covered with bands of different lengths. The feet are yellow.

Golden pheasants mainly inhabits sparse forest shrubs, coniferous and broad-leaved mixed forests and broad-leaved forests up to 3000m. It prefers rocky slopes with high Forest Canopy Closure and open undergrowth, where bushes or bamboo thickets are present, and even farmland environments. There is no obvious preference for the orientation of the hillside. The bird is only found in China, distributed in Qinghai, Gansu, Ningxia, Shaanxi, Henan, Sichuan, Xizang, Hubei, Hunan, Yunnan, Guizhou, and Guangxi. In the Qionglai mountain system of Sichuan, natural hybrids with Lady Amherst's Pheasant are found.

14. 白腹锦鸡 Lady Amherst's Pheasant

鸡形目　Galliformes　雉科　Phasianidae　*Chrysolophus amherstiae*

　　雄鸟体长约130cm，雌鸟体长约60cm的大型雉科鸟类。雌雄两性羽色区别较大。雄鸟前额金属绿色，头顶具一簇红色羽冠，后颈披一白色而具蓝绿色羽缘的扇状羽，形成披肩状；颈部、胸部、上背部、肩部具金属绿色鳞状羽，外缘近黑色，两翼具紫蓝色金属光泽；腹部、两胁白色，向后转为朱红色；白色尾羽甚长，间以黑色横斑，有红色羽饰。雌鸟整体棕褐色，胸部棕栗色具黑色细纹，背部、两翼多黑色和棕黄色横斑。

　　白腹锦鸡在中国见于西藏东南部、四川西南部、贵州西部、广西西部和云南，亦见于邻近的缅甸北部，英国已成功引入野化多年。为典型的昼行性林栖鸟类，白天活动机敏，反应迅速，夜间栖息时反应迟缓，防御力较低。生活在中高海拔常绿阔叶林、针叶林、针阔混交林和竹林等地带，亦见于林缘灌丛、林缘草坡和疏林荒山，夜间栖息选择坡度较平缓、容易起飞、隐蔽性好的高大的乔木栖枝，利于躲避天敌和遮风挡雨。白腹锦鸡主要以草籽及其他植物种子为食。叫声单调，有时会发出响亮而枯燥的似"嘎-嘎嘎"的叫声。

Lady Amherst's Pheasant is a large bird of the Phasianidae family with body length of about 130cm for male and 60cm for female. The male and female plumage are quite different. The male bird has a metallic green forehead, cluster of red feathers on cap, and white fan-shaped feather with bluish green edges on the nape, forming shape of a shawl; the neck, chest, mantle, and shoulders have scaly metallic green feathers with blackish edges; purple-blue metallic luster on both wings; the abdomen and flanks are white, turning vermilion toward back; the white tail feathers are very long, with black bands between them and together with red plumes. The female is generally rufous, chest are chestnut brown with black fine lines, and black and brownish yellow bands on the back and wings.

Lady Amherst's Pheasant is found in China's southeastern Xizang, southwestern Sichuan, western Guizhou, western Guangxi and Yunnan, as well as neighboring country northern Burma. Britain has been successfully introduced this bird into rewilding for many years. These typical diurnal forest-dwelling birds are active and react quickly during the day. They are slow to react and have low defenses at night. They live in areas such as evergreen broad-leaved forests, coniferous forests, coniferous and broad-leaved mixed forests, and bamboo thickets at medium high altitude. It is also found in shrubs and grass slopes of forest edges, as well as sparse forest barren hills. During the night roosting ground is chosen in a flat area where they can easily take off from well-concealed branches of tall trees, as this habitat has the advantage for avoiding natural enemies and sheltering from wind and rain. Lady Amherst's Pheasant mainly feeds on grass seeds and other plant seeds. The call is monotonous, sometimes a loud and boring call like quack, quack-quack.

雌鸟　　摄影：罗平钊

雄鸟　摄影：董磊

15. 黑颈鹤 Black-necked Crane

鹤形目　Gruiformes　鹤科　Gruidae　*Grus nigricollis*

体长约115cm的大型鹤科鸟类。雌雄两性于野外较难区别。头顶裸皮为红色，前额为黑色，眼后具一白斑，成鸟虹膜淡黄色，幼鸟虹膜深褐色，喙部为灰绿色；头颈部多为黑色；胸腹部、背部为灰白色，两翼覆羽灰白色，飞羽沾黑色；尾羽近黑色。黑颈鹤与其他鹤类最显著的区别是本种颈部黑色，无浅色区域，并且是世界上唯一一个繁殖、越冬于高原地区的鹤种，也是1876年由俄国探险家在青海省青海湖畔获取标本后科学命名的最后一种鹤类。

黑颈鹤种群除少数迷鸟[1]外，主要见于青藏高原和云贵高原，分布于海拔从2000米至5000米的高山草甸、湿地、农田等地带。黑颈鹤迁徙途中会从西南山地区域经过和逗留，鸟类环志[2]已经证实了四川若尔盖至贵州草海、青海隆宝滩至云南纳帕海的迁徙线路，两地直线距离均超过800公里。而推测青海西部和西藏北部的黑颈鹤还会一路南下进入雅鲁藏布江河谷，有些还会穿过喜马拉雅山脉，进入不丹等国。

黑颈鹤主要以植物根、茎、叶等为食，大风期间多迎风行走取食，而无风情况下则背向太阳取食，推测与能量消耗策略有关。繁殖期在四川和青海还有见到捕捉小型动物，包括高原林蛙、沙蜥类、红脚鹬、赤麻鸭的幼鸟、高原鼠兔等。黑颈鹤叫声较为单调，常发出高亢响亮的似"尔-咦，尔-尔"的叫声。

[1] 迷鸟：鸟类偶尔因狂风等气候骤然变化，或依随船舶飞行，从平常的栖息区域或正常的迁徙路径，飘零至异地，即为该地区的迷鸟。
[2] 鸟类环志：在鸟类身上佩戴刻有特定标记的金属或塑料环，用以观察研究其活动规律的一种方法。

摄影：唐军

Black-necked Crane is a large bird of the Gruidae family with a body length of about 115cm. It is difficult to distinguish between male and female in the wild. The bare skin on the cap is red, it has black forehead and a white patch behind each of the eyes. Adult bird's iris is pale yellow, while young bird's iris is dark brown, and the bill is grey-green; the head and neck are black; the chest, abdomen and back are grey-white, with grey-white wing coverts and black flight feathers; the tail feathers are blackish. The most significant difference between the Black-necked Crane and other cranes is that this species has a black neck and no pale-colored areas. It is the only crane species in the world that breeds and overwinters in plateau areas. It was also the last species of crane scientifically named by Russian explorers in 1876 after the specimens were obtained from Qinghai Lake in the province.

Except for a few lost birds, the Black-necked Crane is mainly found in the Qinghai-Xizang Plateau and the Yunnan-Guizhou Plateau. It is distributed in alpine meadows, wetlands, farmland, and other areas from 2000m to 5000m. Black-necked Crane will pass through and stay in the Mountainous of Southwest China during their migration. Bird bandings have confirmed the migration route was from Ruo'ergai in Sichuan to Caohai in Guizhou, and Longbaotan in Qinghai to Napahai in Yunnan. The linear distance between the two places exceeds 800 kilometers. It was speculated that the Black-necked Cranes in western Qinghai and northern Xizang will go all the way south into the Yarlung Tsangpo River Valley, and some will pass through the Himalayas and enter Bhutan and other countries.

Black-necked Crane mainly feeds on plant roots, stems, leaves, etc. During strong winds, they tend to walk against the wind to feed; or turn their backs against the sun to feed when there is no wind. It is presumed that this behaviour is related to energy consumption strategies. During the breeding season, small animals were also caught as food in Sichuan and Qinghai, including the Plateau Brown Frog and the sand lizard, Redshank, young bird of Ruddy Shelduck, Plateau Pika,etc.. The call of Black-necked Crane is relatively monotonous, often with high-pitched and loud call of er-huh, er-er.

16. 林沙锥 Wood Snipe

鸻形目 Charadriiformes 鹬科 Scolopacidae *Gallinago nemoricola*

体长约30cm的中型鹬科鸟类。雌雄两性于野外较难区别。成鸟整体羽色暗淡；头部褐色，具较明显皮黄色眉纹，深褐色过眼纹、侧顶纹；喉部棕黄色具深褐色纵纹，胸腹部棕黄色具深褐色纵纹；背部具两道明显的棕黄色纵纹；棕红色尾羽较短，末端白色。

林沙锥在国内见于西藏、云南、四川等地区，为各地罕见夏候鸟。夏季栖于高海拔草地、灌丛、沼泽、池塘等地，冬季迁徙至低山平原的河流、沼泽等地。常单独活动，胆小孤僻，较为惧人。主要以昆虫、昆虫幼虫和小型动物为食。叫声较为单调，有时会发出似撞击石头般的"哒-哒-哒"声。林沙锥在我国分布范围狭窄，数量稀少，加之胆怯惧人，故在分布区内遇见率很低，在四川巴朗山地区中高海拔的多石草甸地带有较低的遇见率，繁殖期有时在天擦亮阶段于空中做炫耀飞行。

Wood Snipe is a medium-sized bird of the Scolopacidae family with a length of about 30cm. It is difficult to distinguish between male and female in the wild. The general plumage of the adult bird is dim; the head is brown, with more obvious yellow eyebrow lines, dark brown eye-stripes, and lateral coronal stripes; brownish yellow throat with dark brown vertical lines, chest and abdomen with dark brown vertical lines; two obvious brownish-yellow stripes on the back; short brownish-red tail feathers with white tips.

Wood Snipe is found in Xizang, Yunnan, Sichuan and other regions in China and is considered rare summer resident in these regions. It inhabits places such as high-altitude grasslands, shrubs, swamps, and ponds in summer, and migrates to rivers and swamps in low mountain plains in winter. It is often alone, timid and unsociable, fearful to human. It mainly feeds on insects, insect larvae, and small animals. The call is relatively monotonous, and sometimes it makes a call of da, da, da like hitting a stone. Wood Snipe is narrowly distributed in China, and the number is small. In addition, they are afraid of people. Therefore, the encounter rate in the distribution area is very low. The medium and high altitude stony meadows in the Balang Mountain area of Sichuan have a low encounter rate. Sometimes, it flies ostentatiously when light first appears in the sky.

摄影：唐军

17. 大紫胸鹦鹉 Derbyan Parakeet

鹦鹉目　Psittaciformes　鹦鹉科　Psittacidae　*Psittacula derbiana*

雄鸟体长约46cm，雌鸟体长约41cm的大型鹦鹉科鸟类。雌雄两性形态有一定的区别。雄鸟头部主要为蓝紫色，眼先具一道黑色斑纹，上喙鲜红色，下喙深色；下颏、喉部及颈侧黑色，枕部绿色；胸腹部蓝紫色；背部、两翼为绿色；尾下覆羽绿色，尾羽蓝绿色。雌鸟似雄鸟，但喙部主要为黑色。

大紫胸鹦鹉在国内见于西藏、云南、四川、广西等地区，国外还见于印度北部，均为当地留鸟。常结群栖于中高海拔的阔叶林、针叶林、针阔混交林等地带。主要以植物果实、种子为食，有时亦捕食昆虫。叫声较为单调，常发出连续而尖锐的"啊，啊"声。

左雌鸟右雄鸟　摄影：李利伟

摄影：董磊

　　The male bird is about 46cm and the female bird is about 41cm in length. There are certain differences in appearances between males and females. The male bird's head is mainly bluish purple, with black markings on the lores, bright red upper mandible, and dark lower mandible; black chin, throat and side of neck, green nape; bluish purple chest and abdomen; green back and wings; green undertail coverts, and bluish green tail feathers. The female bird resembles the male bird, but the bill is mainly black.

　　Derbyan Parakeet is found in Xizang, Yunnan, Sichuan, Guangxi and other regions in China; it is also distributed in northern India. This parakeet is a local resident bird to these regions. It often lives in medium and high altitude broad-leaved forests, coniferous forests, and coniferous and broad-leaved mixed forests. It mainly feed on fruits and seeds and sometimes prey on insects. Its call is relatively monotonous, often with continuous and sharp call of ah, ah.

18. 鹊鹂 Silver Oriole

雀形目　Passeriformes　黄鹂科　Oriolidae　*Oriolus mellianus*

体长约28cm的大型黄鹂科鸟类。雌雄两性于野外区别较大。雄鸟头部黑色，淡黄色虹膜于头部非常明显，喙部铅灰色；颈部、胸部、腹部为灰色；背部灰色，两翼多黑色；尾下覆羽及尾羽正红色。雌鸟似雄鸟，但背部颜色为深灰色，喉部、胸腹部偏白色并具黑褐色纵纹。鹊鹂雌鸟在野外与朱鹂（*Oriolus traillii*）雌鸟有一定的识别难度，但本种背部颜色为灰色，朱鹂雌鸟为红褐色。

鹊鹂在国内仅分布于西南地区，为各地少见夏候鸟。常栖于中低海拔的常绿阔叶林地带。主要以昆虫为食，有时亦吃植物果实、种子。

Silver Oriole is a large bird of the Oriolidae family, with a body length about 28cm, sexually dimorphic. With sharp contrast between the pale-yellow iris and the black head, the male bird has plumbeous grey bill and grey neck, breast, abdomen and back. The wings are mostly black, the undertail coverts and tail feathers are red. The female bird resembles the male, but its back is dark grey, and the throat, chest, and abdomen are white with blackish brown stripes. Silver Oriole females are difficult to distinguish from female Maroon Orioles (*Oriolus traillii*) in the wild, but the back of this species is grey while that of the Maroon Oriole females are reddish brown.

In China, Silver Oriole is only distributes in the southwestern regions and is considered a rare summer resident in such areas. It often inhabits in evergreen broad-leaved forests at low-altitude areas, mainly feeding on insects, and sometimes also on plant fruits and seeds.

雌鸟　摄影：罗平钊

雄鸟　摄影：钟宏英

19. 黑头噪鸦 Sichuan Jay

雀形目　Passeriformes　鸦科　Corvidae　*Perisoreus internigrans*

体长约30cm的小型鸦科鸟类。雌雄两性于野外较难区别，成鸟整体多为灰色，头色深偏黑，嘴黄色分外显眼。

黑头噪鸦为中国狭域分布的特有留鸟，仅见于青藏高原东部高山和亚高山针叶林，已知栖息地散布在四川西部、甘肃南部、青海东南部和西藏东北部，各地理种群多间断分布，四川九寨沟地区靠近该物种的分布中心。它们多在地面取食动物尸体、植物果实和种子、昆虫及其他无脊椎动物等。

摄影：钟宏英

摄影：杨宪伟

Sichuan Jay is a small bird of the Corvidae family with a body length of about 30cm. It is difficult to distinguish between male and female in the wild. Adult birds are mostly grey in plumage, with a darkish black head and an obvious yellow bill.

Sichuan Jay is an endemic resident bird distributed in a narrow range of China. It is only found in alpine and sub-alpine coniferous forests in the eastern part of the Qinghai-Xizang Plateau. Known habitats scatter in western Sichuan, southern Gansu, southeastern Qinghai, and northeastern Xizang; the distributions of different populations are geographically intermittently. Jiuzhaigou Valley of Sichuan is close to the distribution center of this species. They often feed on animal carcasses, fruits, seeds, insects and other invertebrates on the ground.

20. 红腹山雀 Rusty-breasted Tit

雀形目　Passeriformes　山雀科　Paridae　*Poecile davidi*

　　体长约12cm的小型山雀科鸟类。雌雄两性于野外较难区别。成鸟头顶、喉部为黑色，脸颊白色；上体深灰色，下体棕红色；尾羽灰色。

　　红腹山雀常栖于海拔1500m以上的高山针叶林和针阔混交林环境，有时亦见于竹林。除繁殖季外，常成小群活动，夏季主要以昆虫为食，冬季主要以植物果实、种子为食。常发出连续的具金属声的"啾喂喂喂"声。

Rusty-breasted Tit is a small bird of the Paridae family with a body length of about 12cm. It is difficult to distinguish between male and female in the wild. The head and throat of the adult bird are black, and the cheeks are white; the upperpart is dark grey, and the underpart is brownish red; the tail feathers are grey.

Rusty-breasted Tit usually inhabits alpine coniferous forests and coniferous and broad-leaved mixed forests above 1500m, but sometimes it is also seen in bamboo thickets. Out of the breeding season, they often live in small groups. In summer the birds mainly feed on insects, while in winter they mainly feed on fruits and seeds of plants. A continuous metallic call of jiu-wei-wei-wei is often made.

摄影：罗平钊

21. 四川褐头山雀 Sichuan Tit

雀形目　Passeriformes　山雀科　Paridae　*Poecile weigoldicus*

　　体长约12cm的小型山雀科鸟类。雌雄两性于野外较难区别。成鸟头顶、喉部为褐色，脸颊米黄色；上体灰褐色，两翼多灰色，下体浅棕色；尾羽偏灰色。

　　国内见于西藏、云南、四川、青海等地区，为地方性较常见留鸟。夏季常栖于中高海拔的高山针叶林和针阔混交林环境，冬季下至低海拔地区活动。除繁殖季外，常成小群活动，主要以昆虫和植物种子为食。常发出连续的具金属声的"咦喂喂，咦喂喂"声。

　　Sichuan Tit is a small bird of the Paridae family with a body length of about 12cm. It is difficult to distinguish between male and female in the wild. The head and throat of the adult bird are brown, and the cheeks are creamy; the upperparts are greyish brown, the wings are mostly greyish, and the underparts are pale brown, the tail feathers are greyish.

　　It is found in areas such as Xizang, Yunnan, Sichuan, Qinghai, and is a locally common resident where it occurs. It often inhabits alpine coniferous forests and coniferous and broad-leaved mixed forests of middle and high altitude in summer, and moves down to lower altitude in winter. Out of the breeding season, they often move in small groups, mainly feeding on insects and seeds. A continuous metallic call of yi-wei-wei, yi-wei-wei is often made.

摄影：唐军

22. 领雀嘴鹎 Collared Finchbill

雀形目　Passeriformes　鹎科　Pycnonotidae　*Spizixos semitorques*

　　体长约22cm的中型鹎科鸟类。雌雄两性于野外较难区别。体色大体为橄榄绿，喙为象牙白色，厚而短粗，故名"雀嘴"。成鸟头部深灰色，眼周附近黑色，下颏和喉部近黑色，颈部具一细窄浅白色颈环，尾羽末端为黑色。

　　领雀嘴鹎在中国南方中低海拔区域不难见到，也分布于中南半岛北部，为各地留鸟。常栖息在落叶阔叶林、林缘疏林灌丛、河岸灌丛、草坡灌丛，也有少量在村镇稀疏林环境活动。常结成小群在高处枝头活动。主要以植物果实为食，亦捕食昆虫。叫声具弹音，常发出似"啪德儿，啪德儿"声。

Collared Finchbill is a medium-sized bird of the Pycnonotidae family with a body length of about 22cm. It is difficult to distinguish between male and female in the wild. The body plumage is generally olive green; the bill is ivory white, thick, and stubby, hence it is given the name "Finch Bill". The adult has a dark grey head, black orbital area, blackish chin and throat, a narrow pale white neck ring on the neck, and black tips of tail feathers.

Collared Finchbill is not difficult to encounter in the low and middle altitude areas of southern China, and it also distributes in northern Indo-China Peninsula. This finchbill is a resident bird in these regions. It often inhabits deciduous broad-leaved forests, sparse forest shrubs at the forest edge, riparian shrubs, and grass slope shrubs. There are also a small amount of birds live in sparse forests in villages and towns. They often form small flocks and are active high above branches. It mainly feeds on fruits as well as insects. The calls are flicking, and they often make a call like pa-der, pa-der.

摄影：何屹

23. 银脸长尾山雀 Sooty Bushtit

雀形目　Passeriformes　长尾山雀科　Aegithalidae　*Aegithalos fuliginosus*

体长约11cm的小型长尾山雀科鸟类。雌雄两性于野外较难区别。成鸟最主要的识别特点是眼周为灰白色；顶冠纹棕色，侧顶纹、喉部近黑色；颈部具一圈白色颈环，向上环绕至颈侧；上体深褐色；下体胸部近黑色，腹部偏白色；尾羽较长，近黑色。

银脸长尾山雀分布范围非常狭小，为中国西部特有留鸟，仅见于四川西部和东北部、甘肃南部、宁夏北部、湖北西部、重庆北部、陕西秦岭以及邻近的山西南部和河南西部。生活在1000—2000米的中高海拔山地森林，尤好针叶林与栎树混交林、栎树林，也见于低矮乔木和灌丛。除繁殖季节外，常成小群活动，捕食鳞翅目和鞘翅目昆虫的幼虫和其他蠕虫，也取食植物的嫩叶和果实。常发出连续的柔和金属声，似"句，句，句，吱"，群体觅食时叫声略显嘈杂，不难在野外发现和观察。

Sooty Bushtit is a small bird of the Aegithalidae family with a body length of about 11cm. It is difficult to distinguish between male and female in the wild. The main diagnostic feature of adult birds is that the orbital area is greyish white; the coronal stripe is brown, with blackish lateral coronal stripes and throat; the neck has a white collar that wraps up to the side of the neck; The upperparts are dark brown; in the underparts, the chest is almost black, the abdomen is whitish; the tail feathers are long and in blackish color.

Sooty Tit has a very narrow distribution range and is a resident bird endemic to western China. It is only found in western and northeastern Sichuan, southern Gansu, northern Ningxia, western Hubei, northern Chongqing, Qinling Mountains of Shaanxi, as well as neighboring southern Shanxi and western Henan. It lives in middle-to-high altitude mountain forests of 1000m to 2000m, and especially prefers coniferous and oak mixed forests and oak forests, but also found in low trees and shrubs. Out of the breeding season, they often live in small groups, preying on the larvae and other worms of lepidopteran and coleopteran insects, as well as the young leaves and fruits of plants. A continuous soft metallic call like ju-ju-ju-zhi is often made. The calls are slightly noisy when flocks of birds are foraging, so it is not difficult to find and observe them in the wild.

摄影：何屹

24. 峨眉柳莺 Emei Leaf Warbler

雀形目 Passeriformes 柳莺科 Phylloscopidae *Phylloscopus emeiensis*

体长约12cm的大型柳莺科鸟类。雌雄两性于野外较难区别。整体呈绿色；头部眉纹及顶冠纹偏黄色；上体为橄榄绿色，翅上具两道黄色翅斑；下体灰白色。它的暗色侧顶纹较淡，尤其是后部呈暗绿灰色。暗色眼纹边缘更淡，耳羽其他部位边缘更暗。另外，峨眉柳莺最外侧尾羽具有窄而不太清晰的白边，翅稍圆。

峨眉柳莺的已知分布区为陕西南部、四川中部和东南部、云南东北部、贵州东部梵净山和广东东北部。常栖息于中海拔的山地森林环境，偏好下层稠密的落叶阔叶林。常单独或成对活动。主要以昆虫为食。峨眉柳莺的鸣唱清晰且轻柔，稍有颤音，与冠纹柳莺和白斑尾柳莺那种似山雀的鸣唱迥然不同。

Emei Leaf Warbler is a large bird of the Phylloscopidae family with a body length of about 12cm. It is difficult to distinguish between male and female in the wild. Being a largish leaf warbler of green plumage in general, the eyebrow and coronal stripe are yellowish; the upperparts are olive green with two yellow wingbars; the underparts are greyish white. Its dark lateral coronal stripes are pale, and the rear end is especially dark greenish grey. The borders of dark eye-stripes are paler, and the edges of rest of ear coverts are darker. In addition, the outermost tail feathers of Emei Leaf Warbler have narrow and inconspicuous white edges and the wings are slightly rounded.

The known distribution areas of Emei Leaf Warbler are southern Shaanxi, central and southeastern Sichuan, northeastern Yunnan, Fanjing Mountain in eastern Guizhou, and northeastern Guangdong. It often inhabits mountain forests of middle altitude, and prefers dense deciduous broad-leaved forests. Activities are carried out alone or in pairs. And the bird mainly feeds on insects.The song of Emei Leaf Warbler is clear and soft, with a slight trill, which is very different from the tit-like songs of Claudia's Leaf Warbler and Kloss's Leaf Warbler.

摄影：唐军

25. 四川短翅蝗莺 Sichuan Bush Warbler

雀形目　Passeriformes　蝗莺科　Locustellidae　*Locustella chengi*

　　体长约14cm的蝗莺科鸟类。雌雄两性于野外较难区别。整体棕色，无明显特点。头部棕褐色，具不甚明显的浅褐色眉纹，喉部偏白色；上体棕褐色；下体偏褐色；尾羽较长，为暗褐色。在野外与高山短翅蝗莺（*Locustella mandelli*）很难区分，主要依靠鸣声区别。

　　四川短翅蝗莺原为高山短翅蝗莺一亚种，现独立成单独一种，为中国特有鸟种。国内仅见于四川地区，为当地不常见夏候鸟。常栖息于中低海拔的疏林灌丛环境。性隐秘，不易被发现，在野外主要依靠叫声寻找。常单独或成对活动。主要以昆虫和植物种子为食。常发出具金属音的"咦–喂，咦–喂，咦–喂"声。

　　Sichuan Bush Warbler is a bird of the Locustellidae family with a body length of about 14cm. It is difficult to distinguish between male and female in the wild. The plumage is brown in general without obvious characteristics. The rufous head has very inconspicuous pale brown eyebrows, the throat is white; the upperparts are rufous and the underparts are brownish; the tail feathers are longish and dark brown. It is difficult to distinguish between this species and the Russet Bush Warbler (*Locustella mandelli*) in the wild.

　　Sichuan Bush Warbler used to be a subspecies of Russet Bush Warbler and now splited to be a separate species, which is endemic to China. It is only found in Sichuan, being an uncommon summer visitor. The bird often inhabits sparse shrubs of low altitude. It is quite secretive, and difficult to be found; it is mainly searched by calls in the wild. Activities are usually carried out alone or in pairs. The bird mainly feeds on insects and seeds. A metallic call of yi-wei, yi-wei, yi-wei is often made.

摄影：唐军

26. 金额雀鹛 Golden-fronted Fulvetta

雀形目　Passeriformes　幽鹛科　Pellorneida　*Schoeniparus variegaticeps*

　　体长约11cm的幽鹛科鸟类。雌雄两性于野外较难区别。整体淡黄色和前额金黄色是本种鸟类有别于其他雀鹛的重要识别点。头顶灰褐色具白色斑纹，枕部棕红色，脸颊浅黄色，眼侧下处具一明显的黑色点状斑纹；上体灰色，翅上具金黄色、黑色斑纹；下体灰白色；尾羽边缘金黄色。

　　金额雀鹛为中国特有鸟种。国内仅见于四川、广西等地区，为当地少见留鸟。常栖息于中海拔的常绿阔叶林环境，尤好在覆盖苔藓的树干处活动。常单独或成对活动。主要以昆虫为食。常发出较柔弱的具金属音的"哔、哔哔哔、哔"声。

Golden-fronted Fulvetta is a bird of Pellorneidae family with a body length of about 11cm. It is difficult to distinguish between male and female in the wild. The general plumage of pale yellow and golden forehead are important diagnostic features for this species to separate from other fulvettas. The cap is greyish brown with white markings; the nape is brownish red; the cheeks are pale yellow, and there is a clear black spot under each eye; the upperparts are grey, with golden and black wingbars; the underparts are greyish white and the edges of tail feathers are golden.

Golden-fronted Fulvetta is endemic to China, and it is only found in regions like Sichuan and Guangxi, and it is a rare resident in the places it occurs. It normally lives in evergreen broad-leaved forests of middle altitude, and is particularly active in moss-covered trunks. Activities are often carried out alone or in pairs. The bird mainly feeds on insects. A soft metallic beep, beep-beep-beep, beep is often made.

摄影：何屹

27. 红翅噪鹛 Red-winged Laughingthrush

雀形目　Passeriformes　噪鹛科　Leiothrichidae　*Trochalopteron formosum*

又名丽色噪鹛，体长约28cm的噪鹛科鸟类。雌雄两性于野外较难区别。成鸟头顶深灰色具不明显的黑色斑纹，深灰色脸颊外围环绕一圈黑色至眼先及下颏、喉部，喙部黑色；胸部深灰色或黑色，腹部棕黄色；背部棕黄色，两翼初级飞羽、次级飞羽、大覆羽鲜红色，三级飞羽深灰色；尾下覆羽皮黄色，尾羽红色。

红翅噪鹛主要见于中国四川中部和西部、云南北部和广西，以及国外的越南北部。常结群栖息于海拔 900—3000米的山区常绿林、次生林及竹林的地面或近地面处。常成对或成小群活动。主要以昆虫为食。叫声婉转动听，常发出连续的"喂喂咦，喂喂咦"声，在噪鹛科鸟类的歌声进化方面，红翅噪鹛远不及画眉的歌声婉转动听，也不及斑背噪鹛歌声音节多且音调变化较大。

Red-winged Laughingthrush is a bird of the Leiothrichidae family, with a body length of about 28cm, it is difficult to distinguish between male and female in the wild. The adult bird has inconspicuous black markings on its crown, dark grey cheeks with a black circle that reaches to the lores, chin, and throat, and a black bill. It has a dark grey or black breast and brownish yellow belly and back. The pimaries, secondaries, and the greater coverts are bright red; tertials are dark grey. The undertail coverts are fleshy yellow and the tail feathers are red.

Red-winged Laughingthrush is mainly seen in central and western Sichuan, northern Yunnan and Guangxi in China and in northern Vietnam abroad. It usually lives on the ground or near ground level in evergreen forests, secondary forests, and bamboo groves in groups. It lives in pairs or small groups and feeds mainly on insects. It has beautiful and pleasant sounds, often making a consecutive call of we-we-yi, we-we-yi. In terms of bird song evolution in the Leiothrichidae family, the songs of the Red-winged Laughingthrush is far less melodious than the Hwamei, and have less syllables and variations in tones than the Barred Laughingthrush.

摄影：董磊

28. 灰头斑翅鹛 Streaked Barwing

雀形目　Passeriformes　噪鹛科　Leiothrichidae　*Actinodura souliei*

体长约22cm的中小型噪鹛科鸟类。雌雄两性于野外较难区别。成鸟头顶灰色具黄褐色鳞状斑纹，羽毛较蓬松，成明显羽冠，眼先黑色，颊部、耳羽灰色，喙部深灰色；喉部、胸腹部棕黄色具明显的黑色纵纹；背部黄褐色具黑色纵纹，翼上覆羽、尾羽黄褐色具细小而规整的黑色横纹，尾羽具大面积黑色次端斑及不明显的白色端斑。

灰头斑翅鹛见于四川中西部和中南部，云南的中高海拔山区的针阔混交林，尤其喜好在多苔藓的树干中层活动，偶尔发出单调的鸣声，声音略粗哑，为连续的单音节似"归归归归-归归"声，在繁殖季节，雄性表现出比雌性更丰富的鸣声。非繁殖季节常成对或成小群活动，时常混入"鸟浪"活动。主要以苔藓中的昆虫为食。

Streaked Barwing is a small-to-medium-sized bird of the Leiothrichidae family with a body length of about 22cm. It is difficult to distinguish between male and female in the wild. The cap of the adult bird is grey with yellowish brown scaly markings, and the feathers are fluffy and form into conspicuous crown. The lores are black, the cheeks and ear coverts are grey, and the bill is dark grey; the throat, chest and abdomen are brown with obvious black stripes; the back is brown with black stripes. The upperwing coverts and tail feathers are yellowish-brown, with small and regular black bands. The tail feathers are with large areas of black subterminal and inconspicuous white terminal spots.

Streaked Barwing is found in the middle-western and central-southern Sichuan and the coniferous and broad-leaved mixed forests in the middle high-altitude mountainous areas of Yunnan. It is especially fond of activities in the middle part of mossy tree trunks, occasionally making a monotonous call. The sound is slightly rough and dull, with a continuous single syllable like gui-gui-gui-gui, gui-gui. In the breeding season, males show more abundant singing than females. During the non-breeding season, they often move in pairs or small groups, and often mix in "bird waves". The bird mainly feeds on insects in moss.

29. 橙翅噪鹛 Elliot's Laughingthrush

雀形目　Passeriformes　噪鹛科　Leiothrichidae　*Trochalopteron elliotii*

体长约24cm的中型噪鹛科鸟类。雌雄两性于野外较难区别。成鸟整体多为灰褐色；头部灰色，眼周黑色，眼先黑色甚明显；飞羽多为大面积的明显橙黄色，初级飞羽外缘为亮灰色；尾下覆羽为棕红色，尾羽主要为橙黄色，次端斑为灰色，端斑为白色。

橙翅噪鹛为中国特有留鸟，分布于青海、四川、云南、西藏、甘肃、陕西、山西、湖北等区域，小区域见于相邻的印度东北部。栖息地海拔跨度较大，见于常绿阔叶林、针阔混交林、疏林灌丛等环境，尤以长有沙棘、连翘等植物的开阔区域为多，有明显的季节性垂直迁移行为，夏季多在高海拔阴坡活动，冬季则在低海拔阳坡活动。主要以鳞翅目和鞘翅目昆虫为食，也取食草籽和浆果。

Elliot's Laughingthrush is a medium-sized bird of the Leiothrichidae family with a body length of about 24cm. It is difficult to distinguish between male and female in the wild. Adult birds are mostly greyish brown in general; the head is grey and the orbital area and lores are black; the flight feathers form large area of obvious orange-yellow, and the outer edges of the primaries are bright grey. The undertail coverts are brownish red; the tail feathers are mainly orange-yellow, with grey subterminal spots and white terminal spots.

Elliot's Laughingthrush is a resident bird endemic to China, distributing in Qinghai, Sichuan, Yunnan, Xizang, Gansu, Shaanxi, Shanxi, Hubei, etc. from central of the country to southeastern Xizang. It is also found in small areas of northeastern India. Habitats have a large altitude span, and the bird can be found in evergreen broad-leaved forests, coniferous and broad-leaved mixed forests, sparse forest shrubs, etc., especially in open areas with plants such as sea-buckthorn, forsythia, etc. This bird has obvious seasonal vertical migration behavior. It is mostly on the shady slopes of high altitude during summer, and in winter moves to the sunny slopes of low altitude. The birds mainly feed on Lepidoptera and Coleoptera insects and also take grass seeds and berries.

摄影:董磊

摄影:董磊

30. 灰胸薮鹛 Emei Shan Liocichla

雀形目　Passeriformes　噪鹛科　Leiothrichidae　*Liocichla omeiensis*

　　体长约18cm的小型噪鹛科鸟类。雌雄两性于野外较难区别。成鸟头顶灰色具细小的黑色纵纹，眼上具一黑色点状斑纹，眼圈淡黄色，眼周、颊部黄绿色；喉部黄绿色，胸腹部灰色；背部黄绿色；飞羽主要由黄色、红色、黑色组成，形成三种颜色的靓丽翼上斑纹；尾羽主要为灰绿色，具不甚明显的黑色横纹，尾羽末端橙红色或黄色。

　　灰胸薮鹛为中国西南山地特有留鸟，分布区域极其狭窄，仅见于四川中南部的雅安、峨边、屏山、马边等地和云南东北部的山区森林，主要见于常绿阔叶林，海拔范围为500—2200米，并偏向于阳坡。主要以昆虫为食，偶尔也吃植物果实、种子等。叫声婉转，常年可听到雌鸟和雄鸟互相联络的鸣唱，如"居，举句，举句"声，并有多种其他鸣声。

Emei Shan Liocichla is a small bird of the Leiothrichidae family with a body length of about 18cm. It is difficult to distinguish between male and female in the wild. The cap of the adult bird is grey with small black stripes, and there is a black spot above the eye; the eye ring is pale yellow, and lores and cheeks are yellowish green. The throat is yellowish green too, and the chest and abdomen are grey. The back is yellowish green; the primaries are mainly yellow, red, and black, forming a beautiful tri-color wing marking; the tail feathers are mainly greyish green with inconspicuous black stripes, with orange-red or yellow tips.

Emei Shan Liocichla is a resident bird endemic to the mountainous regions of southwest China. Its distribution area is extremely narrow. It can only be found in Ya'an, Ebian, Pingshan and Mabian in central and southern Sichuan as well as mountain forests in northeastern Yunnan. It is mainly found in evergreen broad-leaved forests. The range is 500m to 2200m, and is inclined to the sunny slope. It mainly feeds on insects and occasionally fruits and seeds. The call is tactful; males and females' duet singing can be heard all year round, such as the call of juu, ju-juu, ju-juu, and many other sounds.

摄影：叶昌云

31. 画眉 Hwamei

雀形目　Passeriformes　噪鹛科　Leiothrichidae　*Garrulax canorus*

体长约23cm的中型噪鹛科鸟类。雌雄两性于野外较难区别，多以鸣声区分。成鸟整体黄褐色，头部多为黄褐色，眼周向后延伸至眼后，形成一道明显的白色条纹，如画过的眉纹，英文名因此以中文俗称"画眉"直译；头顶、下颏、喉部、颈部、枕部多为黄褐色，且具细小的深褐色纵纹；背部、腹部多棕黄色，几乎无斑纹；尾下覆羽深棕色，尾上覆羽及尾羽前端为棕褐色，尾羽末端深褐色并具不明显的黄褐色横纹。

画眉生活在中国中部和南部大多数地区、海南岛，以及中南半岛北部，其中分布于中国西南地区的是指名亚种*Garrulax canorus canorus*。常栖于海拔1800米以下的灌木丛和亚热带常绿阔叶林林缘环境，尤好在次生林、茶园等地带活动，繁殖的巢址多缺乏高大的乔木。秋冬季节，画眉则成对或成家族群活动。主要以昆虫为食，有时亦食用植物果实、种子等。叫声非常婉转多变，有多种音节和句法，常发出"喂唧唧-唧喂-唧喂-唧喂啾"声，有时也效仿其他鸟类的叫声。

摄影：罗平钊

Hwamei is a medium-sized bird of the Leiothrichidae family with a body length of about 23cm. It is difficult to distinguish between male and female in the wild, and they are usually distinguished by singing. Adult birds are yellowish brown in general, and their heads are mostly yellowish brown. The orbital area extends to the back of the eyes and with a clear white stripe, like a painted eyebrow pattern. Therefore, the English name of the bird is literally and commonly known as Hwamei (meaning painted eyebrow); the top of the head, chin, throat, neck, and nape are generally yellowish brown with small dark brown stripes; the back and abdomen are mostly brown and with almost no markings; the undertail coverts is dark brown, and the upper tail coverts and base of the tail feather are rufous. The tips of tail feathers are dark brown with inconspicuous yellowish brown bands.

Hwamei lives in most parts of central and southern China, as well as Hainan Island and northern Indo-China Peninsula. The named subspecies *Garrulax canorus canorus* is distributed in southwest China. They often live in shrubs and subtropical evergreen broad-leaved forests below 1800m. They are especially active in secondary forests, tea gardens, and other areas, and the breeding nest sites lack tall trees. In autumn and winter, activities are carried out in pairs or family groups. It mainly feeds on insects, and sometimes also fruits and seeds. The call is very songful and changeable, with a variety of syllables and syntax. It often makes a call of woei-chirp-chirp, chirp-woei, chirp-woei, chirp-woei-chew, sometimes imitating the calls of other birds.

32. 黑额山噪鹛 Snowy-cheeked Laughingthrush

雀形目　Passeriformes　噪鹛科　Leiothrichidae　*Ianthocincla sukatschewi*

　　体长约29cm的中型噪鹛科鸟类。雌雄两性于野外较难区别。成鸟额头、贯眼纹、下颊纹黑色，包围一明显的白色耳羽。上体灰褐色，三级飞羽末端具白色斑点，有时不易看出，初级飞羽外缘蓝灰色；下体多为褐色，两胁沾粉色；尾下覆羽棕黄色，尾羽多为棕色，尾羽末端灰白色。

　　国内见于四川、甘肃等地区，为地方性不常见留鸟。常栖息于中高海拔的亚高山灌丛及林缘地带，尤好在林下灌丛和竹林发达的环境活动。常成对或成小群活动。主要以昆虫为食，有时也吃植物种子、果实。常发出"归咦，归咦"声。

　　Snowy-cheeked Laughingthrush is a medium-sized bird of the Leiothrichidae family with a body length of about 29cm. It is difficult to distinguish between male and female in the wild. Adult birds have black forehead, eye-stripe, and lower cheeks, surrounded by a distinct white ear covert. The upperparts are greyish brown, with white spots on the tip of the tertiary feathers (sometimes not obvious), and the outer edge of the primaries is bluish grey; the underparts are mostly brown, with pink on both flanks; the undertail coverts are brownish yellow, with tail feathers mostly brown, and the tip of tail feathers is greyish white.

　　It is found in Sichuan, Gansu, etc., and is an uncommon resident in places it occurs. It often inhabits sub-alpine shrubs and forest edges at middle and high altitude, especially in the environment where undergrowth shrubs and bamboo thickets are growing. Activities are usually carried out in pairs or small groups. It mainly feeds on insects, and sometimes also seeds and fruits. A call of goo-ee, goo-ee is often made.

摄影：唐军

33. 大噪鹛 Giant Laughingthrush

雀形目　Passeriformes　噪鹛科　Leiothrichidae　*Ianthocincla maximus*

体长约33cm的大型噪鹛科鸟类。雌雄两性于野外较难区别。成鸟头顶深褐色，耳羽、下颏及喉部棕红色；颈部深灰色，上体多棕色且具黑色、白点斑点，初级飞羽灰色；下体棕黄色；尾羽多灰色。

国内见于四川、甘肃、青海、西藏等地区，为地方性常见留鸟。常栖息于中高海拔的亚高山和高山森林灌丛及林缘地带，有时亦见于人工建筑附近。常成对或成小群活动。主要以昆虫和植物种子、果实为食。常发出"咦–谷儿，咦–谷儿，咦–谷儿，咦"声。

Giant Laughingthrush is a large-sized bird of the Leiothrichidae family with a body length of about 33cm. It is difficult to distinguish between male and female in the wild. Adult birds have dark brown cap and reddish brown ear coverts, chin, and throat; the neck is dark grey, the upperparts are brown with black and white spots; the primaries are grey; and the underparts are brownish yellow; the tail feathers are mostly grey.

It is found in areas like Sichuan, Gansu, Qinghai, and Xizang, being a locally common resident in places it occurs. It often inhabits sub-alpine and alpine forest shrubs and forest edges of middle and high altitude; it also can be found near man-made buildings occasionally. Activities are usually carried out in pairs or small groups. The bird mainly feeds on insects as well as seeds and fruits. A call of yi-gu'er, yi-gu'er, yi-gu'er, yi is often made.

摄影：董磊

34. 白点噪鹛 White-speckled Laughingthrush

雀形目　Passeriformes　噪鹛科　Leiothrichidae　*Ianthocincla bieti*

体长约27cm的中型噪鹛科鸟类。雌雄两性于野外较难区别。成鸟眼周白色，头顶、喉部棕褐色；上体、下体棕褐色，具黑色、白色斑纹，初级飞羽外缘近灰色；尾羽棕褐色且端斑白色，次端斑深褐色。

白点噪鹛为狭域分布的中国西南山地特有留鸟，仅分布于从四川西南部木里至云南西北部丽江和德钦这一三角形区域内。常栖息于中高海拔的山地森林、灌丛环境。常单独或成对活动，较少成群活动，冬季与其他噪鹛混群，甚至在村舍的院落和附近的农地觅食。主要以昆虫和植物果实、种子为食。常发出"归，呦归呦"声。

White-speckled Laughingthrush is a medium-sized bird of the Leiothrichidae family with a body length of about 27cm. It is difficult to distinguish between male and female in the wild. Adult birds have white patch around eyes; the cap and the throat are rufous; both upperparts and underparts are brown with black and white markings, and the outer edge of primaries is greyish; the tail feathers are brown colored with white terminal spots and dark brown subterminal spots.

White-speckled Laughingthrush is a narrowly distributed resident bird endemic to the Mountains of Southwest China, only distributing in the triangular area from Muli in southwest Sichuan to Lijiang and Deqin in northwest Yunnan. It often inhabits montane forests and shrubs of high altitude. Activities are usually carried out alone or in pairs, rarely in groups, while in winter they mix with other laughingthrushes, and even search for food in courtyards and farmland near the villages. It mainly feeds on insects as well as fruits and seeds. A call of gui, yo-gui-yo is often made.

摄影：罗平钊

35. 斑背噪鹛 Barred Laughingthrush

雀形目　Passeriformes　噪鹛科　Leiothrichidae　*Ianthocincla lunulata*

体长约26cm的中型噪鹛科鸟类。雌雄两性于野外较难区别。成鸟眼周白色，头顶、喉部棕褐色；上体多为褐色，具黑色虎纹状斑纹，初级飞羽外缘为银灰色；下体多棕褐色且带虎纹状斑纹，前胸沾灰色；尾羽棕褐色且端斑白色，次端斑黑色。

斑背噪鹛为中国特有鸟种，分布于陕西南部、甘肃南部、四川、重庆、湖北西部、宁夏固原六盘山自然保护区。主要栖息于海拔1400—2600米的高山针叶林、针阔混交林、亚热带常绿阔叶林和竹林中，也出现于林缘疏林灌丛、次生林和路边灌丛中。常成对或单独活动，较少成群，冬季则会与其他同域分布的噪鹛混群活动。主要以昆虫和植物果实为食。常发出"古儿，古儿，古儿"声，鸣唱悠扬而响亮。

Barred Laughingthrush is a medium-sized bird of the Leiothrichidae family with a body length of about 26cm. It is difficult to distinguish between male and female in the wild. Adult birds have white patches around the eyes, and the cap and the throat are rufous; the upperparts are mostly brown with black tiger stripes, and the outer edge of primaries is silver grey; the underparts are mostly brown with tiger stripes, the upper chest has a grey tinge; the tail feathers are rufous with white terminal spots and black subterminal spots.

Barred Laughingthrush is endemic to China, distributing in southern Shaanxi, southern Gansu, Sichuan, Chongqing, western Hubei, and Liupanshan Nature Reserve in Guyuan of Ningxia. It mainly inhabits alpine coniferous forests, coniferous and broad-leaved mixed forests, subtropical evergreen broad-leaved forests and bamboo thickets at an altitude of 1400m to 2600m. It also occurs in shrubs of forest edge, secondary forests and shrubs on roadside. Activities are often carried out in pairs or alone, rarely in groups. In winter, they mix with other laughingthrushes distributed sympatrically. It mainly feeds on insects and fruits. A call of gu-er, gu-er, gu-er is often made, while the song is melodious and loud.

36. 棕草鹛 Tibetan Babax

雀形目　Passeriformes　噪鹛科　Leiothrichidae　*Pterorhinus koslowi*

体长约30cm的中型噪鹛科鸟类。雌雄两性于野外较难区别。成鸟整体多为棕色；眼先色深，为灰褐色；喙略下弯，较我国其他两种草鹛——矛纹草鹛和大草鹛弯曲弧度更大；喉部浅灰色；胸腹部常为黄褐色，具不明显的暗棕色纵纹。本种的颜色为明显棕色且纵纹极少，需关注颜色及斑纹特征。

国内见于四川、青海、西藏的澜沧江上游、怒江支流、帕隆藏布江上游等地区。多栖息在高海拔山谷阳坡的林缘地带有柏树、灌丛的环境，常成对或成3—5只的小群活动，主要以昆虫和植物果实、种子为食。叫声单调，为"咕，咕，咕"声，平常一般不发出叫声，受到惊吓后会发出警报，遇到威胁时往高处移动，威胁过于接近后，则下树逃窜或滑翔。

Tibetan Babax is a medium-sized bird of the Leiothrichidae family with a body length of about 30cm. It is relatively difficult to distinguish between male and female in the wild. Adult birds are mostly brown in general; the lores are dark and greyish brown; the bill is slightly curved, which is more curved than the other two species of Babax in China-Chinese Babax and Giant Babax; the throat is pale grey; the chest and abdomen are often yellowish brown, with inconspicuous dark brown stripes. This species is clear brown and has very few stripes. More attention needs to be paid in the features of the color and markings.

It can be found in Sichuan, Qinghai, Xizang, the upper reaches of the Lancang River, the tributaries of the Nu River, and the upper reaches of the Parlung Tsangpo River. It mostly lives in forest edges on the sunny slopes of high-altitude valleys, with cypress trees and shrubs. Activities are mainly carried out in pairs or in small groups of 3 to 5 individuals. It mainly feeds on insects, fruits, and seeds. The call is monotonous, like goo, goo, goo sound. Generally, it remains silent, but after being frightened, and alarming call is made. When threatened, it moves to a higher place; if the threat is too close, it escapes or glides down from the tree.

摄影：董磊

37. 棕噪鹛 Buffy Laughingthrush

雀形目　Passeriformes　噪鹛科　Leiothrichidae　*Pterorhinus berthemyi*

体长约27cm的中型噪鹛科鸟类。雌雄两性于野外较难区别。成鸟整体多为棕色；眼先黑色，眼周裸皮亮蓝色；背部多为棕黄色，两翼棕红色；胸部棕黄色，腹部灰色；尾羽棕红色，尾下覆羽为白色。

棕噪鹛为中国特有鸟种，见于长江以南的中海拔常绿阔叶林、山地灌丛或竹林。常成对或成小群活动，性情羞怯，很少出现在空旷地带。主要以昆虫和植物种子、果实等为食。

Buffy Laughingthrush is a medium-sized bird of the Leiothrichidae family with a body length of 27cm. It is difficult to distinguish between male and female in the wild. Adult birds are mostly brown in general; the lores are black and the bare skin around the eyes is bright blue; the back is mostly brown and the wings are brownish red; the breast is brown and the abdomen is grey; the tail feathers are brown and red and the undertail coverts are white.

Buffy Laughingthrush is endemic to China, found in mid-altitude evergreen broad-leaved forests, mountain shrubs, and bamboo thickets in south of the Yangtze River. Activities are carried out in pairs or small groups. It is shy and rarely appears in open areas. It mainly feeds on insects, seeds, and fruits.

摄影：杨远方

38. 火尾绿鹛 Fire-tailed Myzornis

雀形目　Passeriformes　鸦雀科　Paradoxornithidae　*Myzornis pyrrhoura*

　　体长约12cm的鸦雀科鸟类，是鹛类中体型较小而体色艳丽的种类。雌雄两性在繁殖季节可以区别。如雌鸟羽色比雄鸟偏蓝且少有光泽；雄鸟整体呈现鲜亮的绿色和黄色，而雌鸟则偏暗绿色；雄鸟的喉部和胸部为红色，雌鸟则少有红色或无红色；雄鸟的尾部和翅部有较为鲜艳的红色，而雌鸟对应部位为橙红色；雄鸟尾下覆羽颜色更加鲜艳，头顶部鳞片状斑纹更宽。

　　火尾绿鹛在中国主要见于云南西部、西藏东南部等地区，国外还见于印度东北部、尼泊尔和缅甸北部。常栖于海拔2000米以上的林缘灌丛、竹林等地带，是鹛类适应高海拔栖息的代表物种。常单独活动，非繁殖季节有时结小群活动，并与莺类、雀鹛、太阳鸟等鸟种混群。

　　火尾绿鹛嗜食花蜜，有时也取食昆虫。叫声较为单调，常发出快速的具金属音的"啧啧，啧啧啧，啧啧"联络声，单独觅食时也发出纤细的单音"啧"的叫声。雄性繁殖期领域性较强，鸣叫较多，多为不同单音节的重复，涉及取食、联络、报警等行为，但平常由于高山低氧环境消耗能量，较为稳定的一夫一妻单配制等原因，故鸣唱较少，但鸣唱十分复杂且声波频率较高。

Fire-tailed Myzornis is relatively small bird with gorgeous plumage of the Paradoxornithidae family, with a body length of about 12cm. During the breeding season, it is possible to distinguish between male and female birds. The female has a more bluish and less shiny plumage than the male. The male is bright green and yellow as a whole, while the female is darker green. The male has red throat and breast, while the female has less red or even no red in corresponding parts. The tail and wings of the male bird are in relatively bright red, while the corresponding parts of the female are orange-red. The color of undertail coverts of the male is brighter and the scaly markings on its crown are wider.

In China, Fire-tailed Myzornis is mainly found in western Yunnan, southeastern Xizang, etc.. It is also seen in northeastern India, Nepal, and northern Myanmar. It often lives in shrubs of forest edges and bamboo thickets at an altitude of more than 2000m. It is a representative species of warblers that had successfully adapted to high altitude. It is mostly solitary, sometimes living in small groups mixed with warblers, finches, sunbirds, and other birds during non-breeding seasons.

Fire-tailed Myzornis loves nectar, and it sometimes also feeds on insects. It has relatively monotonous calls, often making rapid metallic tsi-tsi, tsi-tsi-tsi, tsi-tsi contact call and feeble single-syllable call of tsi when foraging alone. The male bird is relatively territorial in breeding season, calls more frequently with repetition of different single syllables mostly, involving foraging, contacting, alarming, and other behaviors. However, due to the energy consumption in the high mountain hypoxia environment and its relatively stable monogamous relationship, in normal times, the singing is relatively less but the song is very complex and at a relatively high frequency.

摄影：董磊

雄鸟　摄影：罗平钊

39. 宝兴鹛雀 Rufous-tailed Babbler

雀形目　Passeriformes　鸦雀科　Paradoxornithidae　*Moupinia poecilotis*

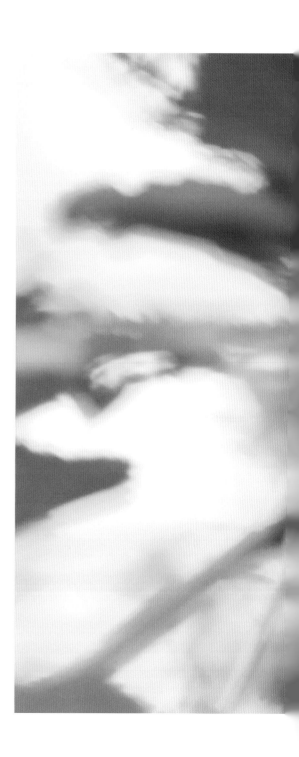

体长约14cm的鸦雀科鸟类。雌雄两性于野外较难区别。成鸟整体为棕褐色；头顶棕褐色，脸部主要为灰色，具不甚明显的偏白色眉纹；喉部主要为白色；胸部为黄褐色，腹部为棕黄色；背部及两翼为棕褐色；尾下覆羽为深棕色，尾羽较长呈棕褐色。

国内见于云南西北部、四川西部地区，为地方性不常见留鸟。常栖息于中高海拔的灌丛地带。常单独或成对活动。主要以昆虫为食。叫声较为单调，常发出具金属音的"句句-句句"声。宝兴鹛雀在云南香格里拉地区的纳帕海和高山植物园地区有较高的遇见率。

Rufous-tailed Babbler is a bird of the Paradoxornithidae family with a body length of about 14cm. It is difficult to distinguish between male and female in the wild. The adult bird is rufous in general; the cap is rufous, the face is mainly grey, with inconspicuous white supercilium; the throat is mainly white; fulvous chested, the abdomen is brownish yellow; it has reddish brown back and wings; the undertail coverts are dark brown, and it has long rufous tail feathers.

It is found in northwestern Yunnan and western Sichuan. It is an uncommon resident bird. It often inhabits shrubland at high altitude. Activities are usually carried out alone or in pairs. It mainly feeds on insects. The call is relatively monotonous, often with metallic call of juu-juu, juu-juu. There is a higher encounter rate of the Rufous-tailed Babbler in Napahai Wetland and Alpine Botanical Gardens in Shangri-La, Yunnan.

摄影：唐军

40. 中华雀鹛 Chinese Fulvetta

雀形目　Passeriformes　鸦雀科　Paradoxornithidae　*Fulvetta striaticollis*

体长约12cm的鸦雀科鸟类。雌雄两性于野外较难区别。成鸟整体为灰褐色；头顶灰色，具不甚明显的黑色纵纹，脸部主要为灰色，具甚不明显的深褐色眉纹，眼先深褐色；喉部主要为污白色；胸腹部及两胁灰白色；背部及两翼主要为棕褐色，初级飞羽具一道不明显的白色翼斑；尾下覆羽为深灰色，尾羽呈棕褐色。

国内见于云南、四川地区，为地方性少见留鸟。常栖息于中高海拔的林缘和杜鹃灌丛地带。常成对或成小群活动。主要以昆虫或植物种子为食。叫声较为单调，常发出具金属音的"哔哔—哔哔"声。

Chinese Fulvetta is a bird of the Paradoxornithidae family with a body length of about 12cm. It is difficult to distinguish between male and female in the wild. The adult bird is greyish brown as a whole; the cap is grey with inconspicuous black stripes; the face is mainly grey with inconspicuous dark brown supercilium, and the lores are dark brown; the throat is mainly dirty white; the chest, abdomen and flanks are grey white; the back and wings are mainly rufous, and the primaries have an inconspicuous white wing spot; the undertail coverts is dark gey, and the tail feathers are brown.

It is found in Yunnan and Sichuan in China and is a rare resident bird. It often inhabits forest edges and rhododendron shrubs at high altitude. It often moves in pairs or small groups. It mainly feeds on insects and plant seeds. The call is relatively monotonous, often with a metallic call of beep-beep, beep-beep.

摄影：何屹

41. 棕头雀鹛 Spectacled Fulvetta

雀形目　Passeriformes　鸦雀科　Paradoxornithidae　*Fulvetta ruficapilla*

　　体长约11cm的鸦雀科鸟类。雌雄两性于野外较难区别。成鸟整体棕色，头顶红棕色，并有黑色边纹延长至颈背。前额近黑色，具一白色眼圈，眉纹深灰色，侧顶纹黑色，脸颊棕色，喙部橙黄色；下颏、喉部白色，具细小的黑色纵纹；胸腹部棕黄色；背部偏灰色，两翼近棕色，外侧初级飞羽具一明显的白色翼斑；尾羽棕黄色。

　　棕头雀鹛在国内见于甘肃、陕西、四川、贵州、云南等地区，在甘肃、陕西南部和四川北部的暖温带地区分布海拔不超过2200米。栖于中海拔的常绿阔叶林、针阔混交林、针叶林和林缘灌丛中，有时也见于农田耕地和村寨附近山坡灌丛中。常成小群活动。主要以昆虫和植物果实、种子为食。叫声较为单调，常发出连续而较为尖锐的"皮皮皮，皮皮皮皮"声。

Spectacled Fulvetta is bird of the Paradoxornithidae family, with a body length of about 11cm. It is difficult to distinguish between male and female in the wild. Adult bird is generally brown, with black lines on two sides of its reddish brown head stretching to the hindneck and mantle. It has blackish grey forehead and supercilium, white eyering, black lateral coronal stripe, brown cheeks, and yellow bill. Fine black lines can be seen on its chin and throat. The breast and abdomen are brownish yellow and the back is greyish. The wings are also brownish, with a distinctive white wingbar on outer primaries and the tail feathers are brownish yellow.

Spectacled Fulvetta can be found in Gansu, Shaanxi, Sichuan, Guizhou, Yunnan, and areas with an altitude less than 2200m in warm temperature zones, such as Gansu, southern Shanxi, and northern Sichuan. It inhabits evergreen broad-leaved forests, coniferous and broad-leaved mixed forests, coniferous forests and bushes on forest edges of middle altitude. They live in small groups, feeding on insects, fruits, and seeds. The song is rather monotonous, usually coming as consecutive sharp pi-pi-pi, pi-pi-pi-pi.

摄影：李利伟

42. 灰头雀鹛 Grey-hooded Fulvetta

雀形目　Passeriformes　鸦雀科　Paradoxornithidae　*Fulvetta cinereiceps*

　　体长约12cm的鸦雀科鸟类。雌雄两性于野外较难区别。成鸟整体为灰褐色；头部银灰色，某些亚种具不甚明显的黑色眉纹；喉部主要为污白色，具不甚明显的褐色斑纹；胸腹部银灰色，两胁黄褐色；背部多为银灰色，两翼主要为棕黄色，初级飞羽具一道较为明显的银灰色翼斑；尾羽深灰色。

　　国内见于青海、四川、甘肃、陕西等地区，为地方性较常见留鸟。常栖息于中高海拔的针阔混交林、竹林、灌丛地带。常成小群活动。主要以昆虫为食。叫声较为单调，常发出具金属声的"哔-哔-哔"声。

Grey-hooded Fulvetta is a bird of the Paradoxornithidae family with a body length of about 12cm. It is difficult to distinguish between male and female in the wild. The adult bird is greyish brown as a whole; the head is silvery grey, and some subspecies have inconspicuous black supercilium; the throat is mainly dirty white with inconspicuous brown markings; chest and abdomen are silvery grey; flanks are yellowish brown. The back is mostly silvery grey; the wings are mainly brownish yellow and the primaries have a more obvious silvery grey wing pattern; the tail feathers are dark grey.

　　It is found in Qinghai, Sichuan, Gansu, Shaanxi, and other regions in China, and is a common resident bird. It often inhabits coniferous and broad-leaved mixed forests, bamboo thickets, and shrubs at high altitude. It is often in small groups. It mainly feeds on insects. The calls are often monotonous, often with metallic call of beep-beep-beep.

摄影：唐军

43. 三趾鸦雀 Three-toed Parrotbill

雀形目　Passeriformes　鸦雀科　Paradoxornithidae　*Cholornis paradoxus*

　　体长约20cm的较大型鸦雀科鸟类。雌雄两性于野外较难区别。成鸟整体为褐色；头部主要为褐色，具明显的黑色眉纹，眼周具一明显白色眼圈，喙部为橙红色；胸腹部褐色；背部多为褐色，初级飞羽沾深褐色；尾羽为灰褐色。第四趾退化严重，只留有残根。

　　三趾鸦雀为中国特有留鸟，只见于四川西部和北部、甘肃和陕西南部地区。常栖息于1500—2700米的中高海拔针阔混交林、竹林、灌丛地带，尤好在较大型的竹林活动。常成对或成小群活动，对人充满好奇，甚至会靠近安静观察。主要以昆虫和植物果实为食。

Three-toed Parrotbill is a largish-sized bird of the Paradoxornithidae family with a body length about 20cm. It is difficult to distinguish between male and female in the wild. Adult birds are brown in general; the head is mainly brown, with obvious black supercilium and white eye ring, and the bill is orange-red; the chest and abdomen are brown; the back is mostly brown. The primaries are dark brown; the tail feathers are greyish brown. The fourth toe is extremely degenerate, leaving only residual root.

Three-toed Parrotbill is a resident bird endemic to China, only found in western and northern Sichuan, Gansu, and southern Shaanxi. It often inhabits the middle-high altitude of 1500m to 2700m in coniferous and broad-leaved mixed forests, bamboo thickets, and shrubs, especially in larger bamboo thickets. They often move in pairs or small groups. They are curious about human, and even observe human closely and quietly. It mainly feeds on insects and fruits.

摄影：何屹

44. 白眶鸦雀 Spectacled Parrotbill

雀形目　Passeriformes　鸦雀科　Paradoxornithidae　*Sinosuthora conspicillata*

体长约14cm的中型鸦雀科鸟类。雌雄两性于野外较难区别。成鸟整体为棕色；头部主要为棕红色，眼周具一甚明显的白色眼圈，喙部为象牙白色，脸颊为棕褐色；胸腹部、背部、尾羽为棕褐色。

国内见于四川、甘肃、陕西等地区，为地方性少见留鸟。常栖息于中高海拔的林缘灌丛、竹林地带。常成对或成小群活动。主要以昆虫为食。叫声较为单调，常发出较为响亮的"丢丢丢-丢丢丢"声。

摄影：何屹

摄影：罗平钊

Spectacled Parrotbill is a medium-sized bird of the Paradoxornithidae family with a body length of about 14cm. It is difficult to distinguish between male and female in the wild. The adult bird is brown as a whole; the head is mainly reddish brown, with a circle of very obvious white eye ring; the bill is ivory white, and the cheeks are rufous; the chest, abdomen, back, and tail feathers are brown.

It is found in Sichuan, Gansu, Shaanxi, and other regions in China, and it is a rare local resident bird. It often inhabits shrubs around forest edge, and bamboo thickets at middle and high altitude. The bird often moves in pairs or in small groups. Mainly feed on insects. The call is relatively monotonous, often with a loud call of dew-dew-dew, dew-dew-dew.

45. 暗色鸦雀 Grey-hooded Parrotbill

雀形目　Passeriformes　鸦雀科　Paradoxornithidae　*Sinosuthora zappeyi*

　　体长约13cm的鸦雀科鸟类，雌雄两性于野外较难区别，羽色无明显季节差异，成鸟整体为灰褐色；头部主要为灰色，眼周具一白色眼圈，喙部较小，为象牙白色；喉部偏白色；胸部偏灰色，腹部及两胁棕黄色；背部、尾羽为棕褐色。

　　暗色鸦雀在中国见于四川、贵州等地区。主要栖息于海拔2300米以上杂有乔木和灌木的浓密竹丛。常单独或成对活动。主要以植物种子、竹子嫩芽和昆虫为食。叫声较为单调，边飞边叫，类似三声的"嘘，嘘嘘"。

Grey-hooded Parrotbill is a bird of the Paradoxornithidae family with a body length of about 13cm. It is difficult to distinguish between male and female in the wild. There is no obvious seasonal difference in plumage, and the whole adult bird is greyish brown; the head is mainly grey, with a white eye ring, the bill is small and ivory white; the throat is whitish; the chest is grey, the abdomen and flanks are brown; the back and tail feathers are rufous.

It is found in Sichuan, Guizhou, and other regions in China. It mainly inhabits dense bamboo thickets with trees and shrubs above 2300m. It is often alone or in pairs. It mainly feeds on plant seeds, bamboo shoots, and insects. The call is relatively monotonous, calling while flying, with the call of shh, shh-shh.

摄影：罗平钊

46. 灰冠鸦雀 Rusty-throated Parrotbill

雀形目　Passeriformes　鸦雀科　Paradoxornithidae　*Sinosuthora przewalskii*

　　体长约14cm的中型鸦雀科鸟类。雌雄两性于野外较难区别。成鸟整体为灰褐色；头顶为深灰色，眼周为明显的棕红色，脸颊为深灰色，喙部较小，为米黄色；喉部在繁殖期为棕红色，胸腹部及两胁深灰色；背部深灰色，两翼为深褐色；尾羽为灰褐色。

　　灰冠鸦雀为中国岷山北部典型的狭域分布特有留鸟，常结小群活动于海拔2400—3500米的开阔针叶林、山区灌丛及竹林中。主要以昆虫为食，也啃食竹子嫩芽。叫声较为单调，常发出具金属音的连续"布吱-布吱"声。

　　Rusty-throated Parrotbill is a medium-sized bird of the Paradoxornithidae family with a body length about 14cm. It is difficult to distinguish between male and female in the wild. The adult bird is greyish brown as a whole; the cap is dark grey, and the orbital area is obviously reddish brown; the cheeks are dark grey, and the bill is small and beige; the throat is brownish-red during the breeding season, and the chest and abdomen are dark grey; the back is dark grey; the wings are dark brown; and the tail feathers are greyish brown.

　　Rusty-throated Parrotbill is a typical resident bird with a narrow distribution in the northern part of Minshan Mountain in China. The flocks of birds are active in open larch forests, coniferous forests, mountain shrubs, and bamboo thickets at an altitude of 2400m to 3500m. It mainly feeds on insects, but also eats bamboo shoots. The call is relatively monotonous, often with a continuous call of buu-jit, buu-jit in a metallic tone.

摄影：唐军

47. 黄额鸦雀 Fulvous Parrotbill

雀形目　Passeriformes　鸦雀科　Paradoxornithidae　*Suthora fulvifrons*

体长约12cm的小型鸦雀科鸟类。雌雄两性于野外较难区别。成鸟整体橙黄色，头顶及前额多橙黄色，具一道显著的黑色侧顶纹，脸颊橙黄色，上喙色深，下喙偏粉色；下颏、喉部橙色；胸腹部多白色；颈部、背部多橙色，两翼近黑褐色，飞羽具橙色羽缘；尾下覆羽白色，尾羽橙色，末端黑色。

黄额鸦雀在中国主要分布于甘肃南部、陕西南部、四川西部和西南部、云南西北部和西部以及西藏东南部，国外还见于尼泊尔、不丹、印度。常栖于中高海拔的针阔混交林、竹林、林缘灌丛等地带，尤好在小型竹林活动。常成对或成小群活动。主要以昆虫为食。叫声较为尖锐，常发出一串快速而具金属音的"吱吱–吱吱吱"声。

Fulvous Parrotbill is a small bird of the Paradoxornithidae family, with a body length about 12cm. It is relatively difficult to distinguish between male and female in the wild. Adult birds are yellowish orange in general. Its crown, forehead, and cheeks are usually yellowish orange, with prominent black lateral coronal strip. The upper mandible is in dark coloration and the lower is slightly pinkish. It has orange chin and throat, almost white breast and abdomen, and mostly orange neck and back. The wings are nearly dark brown with remiges that have orange edges; the undertail coverts are white and the tail feathers are orange with a black tip.

Fulvous Parrotbill mainly distributes in southern Gansu, southern Shaanxi, western and southwestern Sichuan, northwestern and western Yunnan, and southeastern Xizang in China. It is also found in Nepal, Bhutan, and India. It often inhabits coniferous and broad-leaved mixed forests, bamboo thickets, shrubs at forest edges, and other areas at intermediate altitude and particularly favors small bamboo thickets. Activities are usually carried out in pairs or in small groups. The Fulvous Parrotbill feeds mainly on insects. It has relatively sharp calls, often making a series of fast and metallic calls of tsi-tsi, tsi-tsi-tsi.

摄影：唐军

48. 白领凤鹛 White-collared Yuhina

雀形目　Passeriformes　绣眼鸟科　Zosteropidae　*Patayuhina diademata*

　　体长约17cm的绣眼鸟科鸟类，为大型凤鹛。雌雄两性于野外较难区别。成鸟头顶棕褐色，具一明显的白色羽冠，眼先近黑色，喙部橙黄色；下颏黑色，喉部褐色；胸部褐色，腹部米黄色；背部及两翼深褐色，初级飞羽外缘近黑色；尾羽深褐色。

　　白领凤鹛在国外分布于与中国临近的缅甸和越南的北部地区，在中国为西南山地常见林鸟，主要分布于云南、四川、甘肃南部、西藏，也见于湖北、陕西和贵州，为各地留鸟。常结小群活动于中高海拔的阔叶林、针叶林、针阔混交林、竹林等地带，有时亦见于次生林、林缘灌丛、果园等环境。主要以昆虫和植物果实、种子为食。叫声较为单调，为连续而较为尖锐的"喂唧，喂唧，喂唧唧"声，具非常明显的颤音。

摄影：何屹

White-collared Yuhina is a bird of the Zosteropidae family with a body length of about 17cm, being a large Yuhina. It is difficult to distinguish between male and female in the wild. Adult birds have a rufous cap with a distinct white crest. The lores are blackish, and the bill is orange-yellow; the chin is black and the throat is brown; the chest is brown and the abdomen is beige; the back and wings are dark brown, and the outer edges of the primaries are blackish; it has dark brown tail feathers.

Outside China, White-Collared Yuhina distributes in the northern regions of Myanmar and Vietnam, which are close to China. It is a common forest bird in the Mountains of Southwest China. It mainly distributes in Yunnan, Sichuan, southern Gansu, Xizang, and also in Hubei, Shaanxi, and Guizhou. It is a resident bird of these regions. It is often found in small flocks in high-altitude broad-leaved forests, coniferous forests, coniferous and broad-leaved forests, bamboo thickets and other areas, and sometimes in secondary forests, forest edge shrubs, orchards and other environments. It mainly feeds on insects, fruits, and seeds. The call is relatively monotonous, with a continuous and sharp woei-chirp, woei-chirp, woei-chirp-chirp, with a very obvious trill.

49. 滇䴓 Yunnan Nuthatch

雀形目　Passeriformes　䴓科　Sittidae　*Sitta yunnanensis*

体长约11cm的䴓科鸟类。雌雄两性于野外较难区别。成鸟整体为蓝灰色；头顶蓝灰色，头部具一较为显著的白色眉纹，黑色过眼纹自眼后下延至颈侧；喉部、胸腹部为白色或污白色；背部及两翼为蓝灰色，初级飞羽黑褐色；尾羽较短，呈蓝灰色。

国内主要见于云南、四川、西藏等地区，为地方性不常见留鸟。常栖息于中高海拔的山区针叶林、针阔混交林地带。常成对或成小群活动。主要以昆虫为食，有时亦食植物果实或种子等。叫声较为单调，常发出连续的具金属音的"啾啾啾啾啾啾"声。

Yunnan Nuthatch is a bird of the Sittidae family with a body length of about 11cm. It is difficult to distinguish between male and female in the wild. The adult bird is bluish grey in general; the cap is bluish grey, and the head has a more prominent white supercilium. The black eye-stripe extends from behind the eyes to the side of neck; the throat, chest and abdomen are white or dirty white; the back and wings are bluish grey, and the primaries are dark brown; the short tail feathers are bluish grey.

It is mainly found in Yunnan, Sichuan, Xizang, and other regions in China. It is an uncommon resident bird. It often inhabits the coniferous forests and coniferous and broad-leaved mixed forests in the high-altitude mountains. It often moves in pairs or small groups. It mainly feeds on insects, and sometimes also fruits and seeds. The call is relatively monotonous, often with a continuous metallic call of chirp-chirp-chirp-chirp-chirp-chirp.

摄影：王进

50. 白脸䴓 Przevalski's Nuthatch

雀形目　Passeriformes　䴓科　Sittidae　*Sitta przewalskii*

体长约12cm的䴓科鸟类。雌雄两性于野外较难区别。成鸟整体为蓝灰色；头顶黑色；喉部主要为白色；胸腹部为棕黄色，两胁沾棕红色；背部及两翼为蓝灰色，初级飞羽沾深褐色；尾羽较短，呈蓝灰色，中部外缘为白色。

国内主要见于云南、西藏地区，为地方性不常见留鸟。常栖息于中高海拔的山区针叶林、针阔混交林地带。常单独或成对活动。主要以昆虫为食，有时亦食植物种子。叫声较为单调，常发出连续的"句句–句句句"声。

Przevalski's Nuthatch is a bird of the Sittidae family with a body length of about 12cm. It is difficult to distinguish between male and female in the wild. Adult birds are bluish grey in general; the cap is black; the throat is mainly white; the chest and the abdomen are brownish yellow, with brownish red on both flanks; the back and wings are bluish grey, the primaries are dark brown; the tail feathers are bluish grey, and the middle and outer edges are white.

It is mainly found in Yunnan and Xizang, and is an uncommon resident bird. It often inhabits the coniferous forests and coniferous and broad-leaved mixed forests in the high-altitude mountains. It is often alone or in pairs. It mainly feeds on insects, sometimes also on seeds. The call is relatively monotonous, often with continuous call of juu-juu, juu-juu-juu.

摄影：巫嘉伟

51. 巨䴓 Giant Nuthatch

雀形目　Passeriformes　䴓科　Sittidae　*Sitta magna*

　　体长约19cm的大型䴓科鸟类。雌雄两性于野外较难区别。成鸟头顶蓝灰色，头部具一道宽阔的黑色过眼纹，与白色脸颊形成鲜明对比；下颏、喉部、胸腹部米白色；背部蓝灰色，两翼多黑色；尾下覆羽栗色，具白色斑块；尾羽蓝灰色。

　　巨䴓在国内仅分布于西南地区，为各地少见夏候鸟。常栖于中海拔的针叶林和针阔混交林地带。主要以昆虫为食，有时亦取食植物果实、种子，常成对或成小群活动，有时与栗臀䴓（*Sitta nagaensis*）、滇䴓（*Sitta yunnanensis*）混群活动。

Giant Nuthatch is a large bird of the Sittidae family with a body length of about 19cm, and it is relatively difficult to distinguish between male and female in the wild. The adult has bluish grey crown, broad eye-stripe, which contrasts sharply with the white cheek; the chin, throat, breast and abdomen are creamy. The back is bluish grey and the wings are mostly black. The undertail coverts are maroon with white patches and the tail feathers are bluish grey.

In China, Giant Nuthatch only distributes in the southwest regions as a rare summer resident. It often inhabits coniferous forests and coniferous and broad-leaved mixed forests at mid-altitude areas; it mainly feeds on insects and sometimes plant fruits and seeds. It generally lives in pairs or small groups, and sometimes in groups mixed with Chestnut Nuthatch (*Sitta nagaensis*) and Yunnan Nuthatch (*Sitta yunnanensis*).

摄影：何屹

52. 四川旋木雀 Sichuan Treecreeper

雀形目　Passeriformes　旋木雀科　Certhiidae　*Certhia tianquanensis*

体长约14cm的旋木雀科鸟类。雌雄两性于野外较难区别。成鸟整体为灰褐色，喙较其他旋木雀种类更短，头顶灰褐色，具细小的浅色斑纹；眉纹为不明显的白色或皮黄色，脸颊为灰褐色；喉部灰白色，与灰褐色的胸腹部有明显界线；背部灰褐色，具浅色点状斑纹，两翼灰褐色具黄褐色翅斑；尾下覆羽为褐色，尾羽楔形，呈棕褐色，无斑纹。

四川旋木雀为中国西南山地特有留鸟，主要见于四川盆地西缘及北缘山区和陕西秦岭。常栖息于中高海拔的山地环境，尤好在针叶林、针阔混交林活动，作小区域垂直迁移，冬季可见于低海拔阔叶林和针阔混交林，如在四川唐家河国家自然保护区海拔1300—1500米地带较为常见，在四川都江堰城区海拔810米的地方也有发现，多和其他鸟类混群在林间移动和觅食。主要以昆虫为食。叫声较为单调，常发出连续的具金属音的"吱吱吱吱吱吱"声。

Sichuan Treecreeper is a bird of the Certhiidae family with a body length of about 14cm. It is difficult to distinguish between male and female in the wild. Adult birds are greyish brown in general, with a shorter bill than other Certhiidae species; the greyish brown cap has small pale-colored markings; the supercilium are inconspicuous white or yellowish, and the cheeks are greyish brown; the throat is grey and white, having a clear boundary with the greyish brown chest and abdomen; the back is greyish brown with dotted pale-colored markings and the wings are greyish brown with yellowish brown wing patterns; the undertail coverts are brown, and the wedge-shaped tail feathers are rufous without markings.

Sichuan Treecreeper is a resident bird endemic to the mountainous regions of southwest China, mainly found in the mountainous areas of the western and northern margins of the Sichuan Basin and the Qinling Mountains of Shaanxi. It often inhabits high-altitude mountains, especially in coniferous forests and coniferous and broad-leaved mixed forests. It migrates vertically in small regions. In winter, it can be seen in low-altitude broad-leaved forests and coniferous and broad-leaved mixed forests. For example, in the Tangjiahe National Nature Reserve in Sichuan, it is relatively common at an altitude of 1300m to 1500m, and it is also found at an altitude of 810m in the urban area of Dujiangyan, Sichuan. It often mixes with other birds to move and forage in the forest. It mainly feed on insects. The call is relatively monotonous, often with a continuous metallic call of zit, zit, zit, zit, zit, zit.

摄影：董磊

53. 棕背黑头鸫 Kessler's Thrush

雀形目　Passeriformes　鸫科　Turdidae　*Turdus kessleri*

体长约26cm的大型鸫科鸟类。雌雄两性羽色区别明显。雄鸟整体棕、黑两色，头部黑色，下颏、喉部、颈部、枕部皆为黑色，胸部白色，腹部、背部棕黄色，两翼黑色，尾羽黑色。雌鸟浑身浅褐色，头部有灰黑色斑纹，两翼和尾羽黑色，腹部偏棕。

棕背黑头鸫主产于中国西南山地，见于甘肃西部和南部、云南西北部、青海东部和南部及柴达木盆地东缘、四川北部和西部以及西藏东部地区，在国外偶见于印度北部和尼泊尔。棕背黑头鸫在繁殖期主要栖息于海拔3000—4500米的高山针叶林和林线以上的高山灌丛地带，冬季可下至海拔1100米的地带活动。常结小群活动，以昆虫和植物果实、种子为食。叫声较为单调，常发出连续的"句喂句，句句"声。

Kessler's Thrush is a big bird of the Turdidae family, with a body length of about 26cm, sexually dimorphic. The male bird has general plumage of brown and black. The head, chin, throat, neck, and nape are black, while the breast is white, the abdomen and back are brownish yellow. The female bird is generally pale brown, with greyish black lines on its head. The wings and tail are black and the abdomen is brownish.

Kessler's Thrush is mostly found in the southwestern mountains of China, often seen in western and southern Gansu, northwestern Yunnan, eastern Qinghai, eastern margin of the Chaidam Basin, northern and western Sichuan, and eastern Xizang. It is occasionally observed in northern India and Nepal. It mainly inhabits bushes in the areas at an altitude of 3000m to 4500m during the breeding season and comes down to areas of 1100m in winter. Kessler's Thrush lives in small groups and feeds on insects, fruits, and seeds. The call is rather monotonous, usually coming as consecutive sharp ju-we-ju, ju-ju.

雄鸟　　摄影：董磊

54. 宝兴歌鸫 Chinese Thrush

雀形目　Passeriformes　鸫科　Turdidae　*Turdus mupinensis*

　　体长约22cm的中型鸫科鸟类。雌雄两性于野外较难区别。成鸟整体为橄榄绿色；头顶为橄榄绿色，浅色脸颊具两个黑色的斑纹，眼先为偏白色，喙部近黑色；胸腹部为浅褐色，密布黑色的圆形斑纹；背部多为橄榄绿色，翼上具两道不甚明显的浅褐色翼斑，初级飞羽沾棕褐色；尾羽为灰褐色。

　　该种模式产地[①]在四川省宝兴县，故名。宝兴歌鸫为中国特有鸟类，根据形态学研究，亦有科学家认为其是单型种。喜好针叶林和针阔混交林，多单独或成对在林下灌丛和地面觅食。宝兴歌鸫在我国西南山地低海拔春秋季过境期间有一定的遇见率，如成都、昆明主城区绿化较好的公园等。

[①] 模式产地：指物种定名的时候用来定名的原始标本产地。

摄影：何屹

Chinese Thrush is a medium-sized bird of the Turdidae family with a body length of about 22cm. It is difficult to distinguish between male and female in the wild. The adult bird is olive green in general; the cap is olive green, and the pale-colored cheeks have two black spots; the lores are whitish, and the bill is blackish; the chest and abdomen are pale brown with dense black circular markings; the back is mostly olive green, with two inconspicuous pale brown wing spots on the wings; the primaries are rufous; the tail feathers are greyish brown.

The type locality of this bird is Baoxing County, Sichuan Province, hence the name. It is endemic to China. According to morphological research, some scientists believe that Chinese Thrush is a monotypic species. It prefers coniferous forests and coniferous and broad-leaved mixed forests, and is mostly single or in pairs and forages in understory shrubs or on the ground. During the migration seasons of spring and autumn, at low altitude in the mountainous regions of southwest China, there is a certain encounter rate of the Chinese Thrush, such as in green parks of the main urban areas of Chengdu and Kunming.

55. 棕头歌鸲 Rufous-headed Robin

雀形目　Passeriformes　鹟科　Muscicapidae　*Larvivora ruficeps*

　　体长约14cm的鹟科鸟类。雌雄两性形态区别较明显。雄鸟头顶至枕部为橙红色，下颏、喉部为白色，鸣唱时喉部向外鼓起，非常醒目，由眼、嘴角和颈侧三点形成一块黑色的三角区域，与黑灰色渐变的胸带相连接，两胁与之相接，也呈黑灰色，在光线下显露白色鳞片状斑纹；背部及两翼深灰色；腹部及尾下覆羽白色；尾羽中央深灰色，两侧橙红色，末端具黑色端斑。雌鸟整体褐色，下体褐色较浅。虹膜红褐色；嘴黑色；跗跖和趾肉褐色。

　　棕头歌鸲在中国四川北部岷山、陕西南部秦岭一带繁殖，迁徙经过云南往更南的东南亚国家，如马来西亚、柬埔寨等国越冬。繁殖期偏好栖于海拔2350—2700米的针阔混交林和灌丛环境。1907年，西方的研究人员根据在秦岭太白山采集的3号标本为它命名。1986年5月28日，在四川平武县王朗自然保护区的森林中采集到一只成年雄性，为四川省第一笔记录，采集时见其活动于海拔2400米处的生有灌丛的针阔混交林地面，鸣声洪亮多韵，悠扬婉转，十分动听。食性分析发现，所食除一条蚯蚓外，全为植物碎片。

Rufous-headed Robin is a bird of the Muscicapidae family, with a body length of about 14cm, sexual dimorphic. The male bird has orange plumage from the crown to nape and white on the chin and throat, which swells conspicuously when singing. A black triangle is formed by three parts on the bird's face – the eye, base of the bill, and one side of the neck. This triangular area connects with the gradient black breast band, reaching the flanks that are also gradient black and show white scaly stripes under the sunlight. The back and wings are darkish grey while the belly and vent are white. The center of the tail feathers is darkish grey, orange red on two sides, and black terminal spots. The female bird is generally brown, with pale brown belly. The iris is reddish brown, the bill is black, and the tarsus and toes are fleshy brown.

Rufous-headed Robin breeds in the area of Minshan Mountains (northern Sichuan) and Qingling Mountains (southern Shaanxi), migrating through Yunnan to Southeast Asian countries such as Malaysia and Cambodia in winter. It favors bushes and coniferous and broad-leaved mixed forests at an altitude of 2350m to 2700m for breeding. In 1907, western researcher named this specie based on No.3 specimen collected from Mount Taibai in Qinling Mountains. On May 28th, 1986, a specimen of an adult male individual was collected in the forest of Wanglang Nature Reserve in Pingwu, Sichuan, which was the first record in Sichuan Province. It was seen on bushy ground in coniferous and broad-leaved mixed forests at an altitude of 2400m, singing in a pleasant voice that are very sonorous, melodious, and glamorous. The feeding habit analysis showed apart from an earthworm, the rest of the bird's diet is plants.

雌鸟　摄影：JAMES EATON

雄鸟　摄影：PETE MORRIS

56. 黑喉歌鸲 Blackthroat

雀形目　Passeriformes　鹟科　Muscicapidae　*Calliope obscura*

体长约14cm的鹟科鸟类。雌雄两性羽色区别较大。雄鸟头顶暗灰色，眼先及脸颊黑色；喉部、胸部黑色，腹部白色；背部及两翼暗灰色，初级飞羽沾深褐色；中央尾羽多为黑色，外侧尾羽为白色，尾端为黑色。雌鸟整体灰褐色，雄鸟体羽黑色及暗灰色的部分在雌鸟身上被褐色取代。

黑喉歌鸲繁殖于中国陕西南部、甘肃东南部及四川中部和其北部三省相邻的狭小区域内，迁徙时曾见于云南东南部和泰国极北部。黑喉歌鸲迁徙经过四川，在城市校园和周边山区的遇见率极低，而在繁殖地秦岭有相对稳定的遇见率。繁殖前期，雄鸟于竹林间发出急促而响亮的鸣唱，类似具金属音的"唧-喂喂"声，雌鸟基本不鸣叫，或以极低声音鸣叫。

雌鸟　摄影：唐军

Blackthroat is a bird of the Muscicapidae family with a body length of about 14cm. The male and female plumage are quite different. The male has a dark grey cap, black lores and cheeks, a black throat, chest, white abdomen; it has dark grey back and wings, dark brown in the primaries; the central tail feathers are blackish, and the outer tail feathers are white, with black tips. The general plumage of female bird is greyish brown, black and dark grey parts of the male plumage are replaced by brown.

Blackthroat breeds in a small area adjacent to the three provinces of southern Shaanxi, southeastern Gansu, and central and northern Sichuan, and has been seen in southeastern Yunnan and northern Thailand during migration. The Blackthroat migrates through Sichuan, and has a very low encounter rate in urban campuses and surrounding mountainous areas, while in the breeding area of Qinling it is relatively stable. In the early stage of reproduction, the male birds sing quickly and loudly among the bamboo thickets, with a metallic call similar to chirp-woei-woei, the female birds call at little or very low voices.

雄鸟　摄影：何屹

57. 金胸歌鸲 Firethroat

雀形目 Passeriformes 鹟科 Muscicapidae *Calliope pectardens*

体长约14cm的鹟科鸟类。雌雄两性羽色区别较大。雄鸟头顶至后颈深灰色，具细小的黑色斑纹，眼周及脸颊黑色，喙部黑色，颈侧具有较为明显的白色斑块；下颌、喉部、胸部为靓丽的橙红色，腹部皮黄色；背部深灰色，两翼深灰色，飞羽沾黑色；尾下覆羽皮黄色，尾羽黑色。雌鸟整体灰褐色，雄鸟体羽黑色、橙红色、深灰色的部分在雌鸟身上被褐色取代。金胸歌鸲雄鸟在野外与栗腹歌鸲（*Larvivora brunnea*）雄鸟较为相似，但本种颈侧具有较为明显的白色斑块，且无白色眉纹，喉部、胸部为橙红色而非棕黄色。

金胸歌鸲在国内见于四川、陕西等地区，为各地区少见夏候鸟。常栖于中高海拔的针阔混交林、竹林、林缘灌丛地带。常单独活动。主要以昆虫为食。叫声婉转动听，常发出连续的"句句句喂－咦咦句句"声。金胸歌鸲在我国四川巴朗山地区有一定的遇见率。

Firethroat is a bird of the Muscicapidae family, with a body length of about 14cm, sexually dimorphic. The male bird has fine black lines from the crown to the nape, black orbital areas and cheeks, and a black bill. Distinctive white spots can be seen on two sides of the neck. The chin, throat and breast are in bright tangerine while the abdomen is fleshy yellow. Both the back and wings are dark grey, with some blackness on remiges. The undertail coverts are fleshy yellow and the tail feathers are black. The female bird is greyish brown in genera; the black, tangerine, and dark grey plumage on corresponding part of the male is replaced with brown on the female. Males of the Firethroat and the Indian Blue Robin (*Larvivora brunnea*) closely resembles each other in the wild, but the former has distinctive white spot at two sides of its neck and is lack of the white supercilium, aside from that the throat and breast are in tangerine rather than brownish yellow.

In China, Firethroat can be seen in areas such as Sichuan and Shaanxi, as a rare summer resident. It is usually solitary. It inhabits coniferous and broad-leaved mixed forests, bamboo groves, and bushes of forest edges and feeds mostly on insects. It sings melodiously, often making calls in consecutive ju-ju-ju-we, yi-yi-ju-ju. There is a certain encounter rate of the Fire throat in Balang Mountain area in Sichuan, China.

雌鸟　摄影：罗平钊

雄鸟　摄影：何屹

58. 栗背岩鹨 Maroon-backed Accentor

雀形目 Passeriformes　岩鹨科 Prunellidae　*Prunella immaculata*

体长约14cm的小型岩鹨科鸟类。雌雄两性于野外较难区别。成鸟整体蓝灰色，头部多蓝灰色，眼周黑色，喙部深灰色，虹膜淡黄色；下颏、喉部、颈部、枕部、胸部深灰色；腹部、尾下覆羽棕栗色；背部、两翼多栗红色，初级飞羽羽缘具灰色翼斑；尾羽深灰色。

栗背岩鹨见于中国西藏东南部、青海南部、甘肃南部、四川北部及西部、云南北部及西部、陕西南部等地的中高海拔的针叶林林下、林缘灌丛等地带，冬季下迁至植被较好的针阔混交林。常单独或成对活动。主要以昆虫、植物种子为食。叫声非常微弱而尖利，为单音节金属音声。

摄影：罗平钊

摄影：何屹

Maroon-backed Accentor is a small bird of the Prunellidae family with a body length of about14cm, and it is relatively difficult to distinguish between male and female in the wild. Adult birds have greyish blue plumage in general, with a greyish blue crown, black orbital area, dark grey bill and pale-yellow iris. The chin, throat, neck, nape, and breast are dark grey; the abdomen and undertail coverts are chestnut brown; the back and two wings are chestnut red with primaries that have grey wing patches on edges and the tail feathers are dark grey.

Maroon-backed Accentor can be seen at the understory of coniferous forests and shrubs at forest edges in the middle and high altitude areas such as southeastern Xizang, southern Qinghai, southern Gansu, northern and western Sichuan, northern and western Yunnan, and southern Shaanxi. It migrates down to coniferous and broad-leaved mixed forests with better vegetation in winter. It often lives alone or in pairs and mainly feeds on insects and plant seeds. It makes very feeble but sharp calls, with single-syllabled metallic sound.

59. 拟大朱雀 Streaked Rosefinch

雀形目　Passeriformes　燕雀科　Fringillidae　*Carpodacus rubicilloides*

Streaked Rosefinch is a bird of the Fringillidae family with a body length of about 19cm, sexually dimorphic. The male bird is red in general. It has a red head with fine small white spots on the crown, a dark grey bill, red chin and throat, a pink breast, and a abdomen with white spots. Its back is reddish brown with many dark brown stripes and the two wings are usually dark brown. The undertail coverts are pink and the tail feathers are dark brown. The female bird is dark brown in general. With dark brown stripes, the breast, abdomen, and back are yellowish brown and the tail feathers are dark brown.

Streaked Rosefinch is mainly found in Xizang, Gansu, Sichuan, Qinghai, Yunnan, and other regions in the western part of China, as well as in neighboring countries and areas like Uzbekistan, northern India, and Pakistan. It lives alone or in small groups in high-altitude alpine flowstone slope, sparse vegetation areas, and other habitats. It can be seen near the human settlement at an altitude of about 3600m in April and May. It gradually migrates to higher altitude areas after June, up to 4300m to 4500m. The Streaked Rosefinch mainly feeds on the fruits of plants in its habitat and also eats insect larvae during the breeding period. It has a relatively sharp call, often making a metallic we, wegee call.

体长约19cm的燕雀科鸟类。雌雄两性羽色区别较大。雄鸟整体正红色，头部红色，具细小的白色点状斑纹，喙部深灰色；下颏、喉部红色，胸腹部粉红色，皆具白色点状斑纹；背部红褐色，具较多的深褐色纵纹，两翼多深褐色；尾下覆羽粉红色，尾羽深褐色。雌鸟整体深褐色，胸腹部及背部棕黄色，具深褐色纵纹；尾羽深褐色。

拟大朱雀在中国主要见于西部的西藏、甘肃、四川、青海、云南等地，国外还见于乌兹别克斯坦、印度北部、巴基斯坦等国。常单独或成小群活动于中高海拔的流石滩、稀疏树丛等地带，每年4月到5月，可见于海拔3600米左右的人类居住环境附近，6月以后逐渐向高海拔区域迁移，可达4300—4500米。拟大朱雀主要采食栖息地中植物果实，育雏期间也取食昆虫幼虫。叫声较为尖锐，常发出具金属音的"喂，喂叽"声。

雌鸟　摄影：董磊

雄鸟　摄影：董磊

60. 曙红朱雀 Pink-rumped Rosefinch

雀形目　Passeriformes　燕雀科　Fringillidae　*Carpodacus waltoni*

  体长约13cm的燕雀科鸟类。雌雄两性羽色区别较大。雄鸟整体为红色，头顶红褐色，具黑色纵纹，眉纹粉红色，过眼纹深褐色，脸颊为粉红色；下颏及喉部红色；胸腹部及两胁红色且少斑纹；背部红褐色，具深褐色纵纹，两翼深褐色；尾下覆羽红色，尾羽深褐色。雌鸟整体灰褐色，胸腹部及背部多纵纹。雏鸟及幼鸟羽色与雌鸟类似，性别不易区分。

  曙红朱雀在国内见于四川、青海、西藏等地区。常栖息于中高海拔的高山针叶林、针阔混交林地带，有时亦见于林缘疏林、灌丛草地、谷地农田等环境。常成对或成小群活动，有垂直迁徙习性。主要以草籽和其他植物种子为食，繁殖期也取食动物性食物，主要为昆虫幼虫。叫声较为单调，有时会发出具金属音的"呲呲呲–呲呲呲"声。

  Pink-rumped Rosefinch is a bird of the Fringillidae family with a body length of about 13 cm. The male and female plumage are quite different. The male bird is red in general; reddish-brown with black stripes on the cap, pink eyebrows, dark brown eye-stripe, and pink cheeks; red chin and throat; the chest, abdomen and flanks are red with few markings; the back is reddish brown with deep brown longitudinal stripes, dark brown on both wings; red undertail coverts, dark brown tail feathers. The female bird is greyish brown in general, with many stripes on the chest, abdomen, and back. The plumage of chicks and young birds is similar to females, and the sexes are not easy to distinguish.

  It can be seen in Sichuan, Qinghai, and Xizang, etc. in China. It often inhabits high-altitude alpine coniferous forests and coniferous and broad-leaved mixed forests. Sometimes it is also found in forest edges, shrub grassland, valley farmland, and other areas. They often move in pairs or small groups and have the habit of vertical migration. It mainly feeds on grass seeds and other plant seeds. It also feeds on animal food during the breeding season, mainly insect larvae. The call is relatively monotonous, sometimes with a metallic call of tse-tse-tse, tse-tse-tse.

上雄鸟下雌鸟　　摄影：钟宏英

61. 淡腹点翅朱雀 Sharpe's Rosefinch

雀形目　Passeriformes　燕雀科　Fringillidae　*Carpodacus verreauxii*

　　体长约15cm的燕雀科鸟类。雌雄两性羽色区别较大。雄鸟整体紫红色，头顶深紫色，眉纹粉红色，眼后及脸颊深紫色，喙部深灰色；下颏、喉部粉红色，具白色点状斑纹；胸腹部粉红色；背部及两翼多深紫色，大覆羽具较为明显的粉色翼斑；尾下覆羽粉红色，尾羽深紫色。雌鸟整体深褐色；眉纹棕黄色，脸颊深褐色；胸腹部及背部棕黄色，具深褐色纵纹，翼上具黄褐色翼斑；尾羽深褐色。淡腹点翅朱雀在野外与其他紫红色朱雀较为相似，但本种具显著的粉色（雄鸟）及黄褐色（雌鸟）翼斑。

　　淡腹点翅朱雀在国内主要见于四川、云南等地区，为当地不常见留鸟。常栖于中高海拔的针叶林、针阔混交林、林缘灌丛等地带。常成对或成小群活动。主要以昆虫和植物种子、果实为食。叫声较为尖锐，常发出两声一度的具金属音的"吱喂-吱喂"声。淡腹点翅朱雀在我国云南纳帕海地区有一定的遇见率。

雌鸟　摄影：唐军

雄鸟　摄影：唐军

Sharpe's Rosefinch is a bird of the Fringillidae family with a body length of about 15cm. The plumage of male and female is different. The male bird is reddish purple in general. It has dark purple plumage on its crown, ear-coverts and cheek, pink supercilium and dark grey bill. With white spots, the chin and throat are pink, as well as the breast and abdomen. Its back and wings are mainly purple, and the greater coverts have conspicuous pink wingbars; the undertail coverts are pink and the tail feathers are dark purple. The female bird is dark brown in general. It has brownish yellow supercilium and dark brown cheek. With dark brown stripes, the breast, abdomen and back are yellowish brown. The wings have yellowish brown patches and the tail feathers are dark brown. The Sharpe's Rosefinch resembles other purplish rosefinches in the wild, but this species has prominent pink (male) and tawny (female) wingbars.

Sharpe's Rosefinch is mainly seen in Sichuan, Yunnan, etc. in China as an uncommon resident bird. It often inhabits coniferous forests, coniferous and broad-leaved mixed forests, shrubs at forest edge, and other areas at mid-high altitude. It tends to live in pairs or in small group and mainly feeds on insects, plant seeds, and fruits. It has a relatively sharp call, often making two-syllabled, monotone, and metallic call of geewe, geewe. There is a certain encounter rate of the Sharpe's Rosefinch in the Napahai area of Yunnan, China.

62. 斑翅朱雀 Three-banded Rosefinch

雀形目 Passeriformes 燕雀科 Fringillidae *Carpodacus trifasciatus*

体长约18cm的燕雀科鸟类。雌雄两性羽色区别较大。雄鸟整体为红色，头顶红色，前额白色，脸颊及周边白色，颈侧红色；下颏及喉部具白色斑纹；胸部红色，腹部及两胁沾灰白色；背部为红色，两翼深褐色，具两道白色翼斑；尾下覆羽灰色，尾羽近黑色。雌鸟整体灰褐色，雄鸟身体红色的部分在雌鸟身上均被灰褐色取代。

斑翅朱雀在国内见于甘肃、四川、云南等地区，为地方性不常见留鸟。常栖息于中高海拔的高山针叶林、针阔混交林，有时亦见于高山灌丛、草甸和砾石地带。常单独或成对活动。主要以草籽和其他植物种子为食。叫声较为单调，常发出具金属音的"呲呲-呲呲呲"声。

雌鸟　摄影：董磊

雄鸟　摄影：何屹

Three-banded Rosefinch is a bird of the Fringillidae family with a body length of about 18cm. The male and female plumage are quite different. The male bird is red in general, with a red cap, a white forehead, white periphery and cheeks; the side of neck are red; the markings on the chin and throat are white; red chest, greyish white on the abdomen and flanks; red back, dark brown wings with two white wing spots; grey undertail covert, blackish tail feathers. The female bird is greyish brown in general, and the red part of the male bird's body is replaced by greyish brown in that of the female.

Three-banded Rosefinch is seen in Gansu, Sichuan, Yunnan, ect. in China as an uncommon resident bird. It often inhabits high-altitude alpine coniferous forests, coniferous and broad-leaved mixed forests, and sometimes also found in alpine shrubs, meadows, and gravel areas. It is often alone or in pairs. It mainly feeds on grass seeds and other plant seeds. The call is relatively monotonous, often with a metallic call of tse-tse, tse-tse-tse.

63. 白眉朱雀 Chinese White-browed Rosefinch

雀形目　Passeriformes　燕雀科　Fringillidae　*Carpodacus dubius*

体长约17cm的燕雀科鸟类。雌雄两性羽色区别较大。雄鸟整体为红色，头顶深褐色，前额向后延伸至枕部的眉纹为白色，眼周及脸颊为红色；下颌及喉部为红色，具细小的白色斑纹；胸腹部及两胁红色且沾细小的白色斑纹；背部及两翼深褐色，具黑色斑纹；尾下覆羽灰色，尾羽深褐色。雌鸟整体灰褐色，雄鸟身体的红色部分在雌鸟身上均被灰褐色取代，胸腹部多纵纹。

白眉朱雀在国内见于甘肃、青海、四川等地区，为地方性较常见留鸟。常栖息于中高海拔的高山针叶林、针阔混交林，有时亦见于林缘疏林、高山灌丛、草甸等地带。常成对或成小群活动。主要以草籽和一些植物的果实、种子为食。叫声较为单调，常发出羊一般的"咩咩"声。

Chinese White-browed Rosefinch is a bird of Fringillidae family with a body length of about 17cm. The male and female plumage are quite different. The male bird is red in general, with dark brown on the cap, white lines extending back from the forehead to the eyebrow, red orbital areas and cheeks, and red with small white markings on the chin and throat, chest, abdomen, and flanks. The back and wings are dark brown with black markings; the undertail covert is grey and the tail feathers are dark brown. The female bird is greyish brown in general; the red part of the male bird's body is replaced by greyish brown in the female's, and the chest and abdomen have many stripes.

It is found in Gansu, Qinghai, Sichuan, and other regions in China and is a relatively common resident bird. It often inhabits high-altitude alpine coniferous forests, coniferous and broad-leaved mixed forests, and sometimes also found in forest edges, sparse forests, alpine shrubs, meadows,etc.. Activities are usually carried out in pairs or in small groups. It mainly feeds on grass seeds, and fruits and seeds of some plants. The call is relatively monotonous, often making a sheep-like call of baa.

雌鸟　摄影：董磊

雄鸟　摄影：董磊

64. 蓝鹀 Slaty Bunting

雀形目　Passeriformes　鹀科　Emberizidae　*Emberiza siemsseni*

　　体长约14cm的鹀科鸟类。雌雄两性羽色区别较大。雄鸟整体为蓝灰色，头部蓝灰色，眼先深褐色，喙部深灰色；下颏、喉部、胸部及两胁蓝灰色，下腹部至尾下覆羽为白色；背部及两翼多为蓝灰色，翼上沾深褐色；尾羽蓝灰色。雌鸟整体棕褐色，头部棕红色，胸腹部及背部多棕褐色，下腹部为白色，两翼及尾羽深褐色。

　　蓝鹀在国内见于四川、陕西、甘肃、江西等地，为地方性不常见留鸟，有随气候变化作垂直迁移的习性。常栖息于中低海拔的针阔混交林、林缘灌丛等地带，非繁殖季有时亦见于村落附近的林缘灌丛、草地、果园等环境。常单独或成对活动。主要以草籽和其他植物种子为食。叫声较为单调，雄鸟在繁殖期有时会发出具金属音的"呲呲–呲呲–吱吱吱"的鸣唱声。

　　Slaty Bunting is a bird of the Emberizidae family with a body length of about 14cm. The male and female plumage are quite different. The male bird is bluish grey in general, with a bluish grey head, dark brown lores, dark grey bill; bluish grey chin, throat, chest, and flanks; white lower abdomen and undertail coverts, from the abdomen to the vent; back and wings mostly bluish grey, a bit of dark brown can be seen on wings; bluish grey tail feathers. The female is rufous in general, with a brown-red head, brown breasts, abdomen, and back, white lower abdomen, dark brown wings, and tail feathers.

　　It is found in Sichuan, Shaanxi, Gansu, Jiangxi, and etc. in China. It is an endemic and uncommon resident bird and has the habit of vertical migration due to climate change. It often inhabits middle and low altitude areas such as coniferous and broad-leaved mixed forests and forest edge shrubs, etc. Sometimes it is also seen in forest edge shrubs, grasslands, orchards near villages during non-breeding season. It is often alone or in pairs. It mainly feeds on grass seeds and other plant seeds. The call is relatively monotonous, and the male bird sometimes sings with metallic calls of tse-tse, tse-tse, chirp-chirp-chirp during the breeding season.

雄鸟　摄影：何屹

雌鸟　摄影：董磊

附录一　西南山地观鸟线路攻略

唐　军
西南山地SWILD执行董事
ChinaBirdTour创始人

"西南山地"地理上的概念涵盖了几乎整个中国西南的横断山脉以及周边一带，而西南山地的生物多样性景观影像库（www.swild.cn）更是覆盖了青海、西藏、四川、甘肃、陕西、云南、贵州等七个省（自治区）。中国西南山地的生态环境优势不言而喻，这决定了它是中国鸟类资源最为丰富的地方，云南和四川位居中国鸟类资源最为丰富的省份前两位，同时中国76种特有鸟种（郑光美，2011）①中就有60种在中国西南山地区域有分布，特别是四川，特有鸟种数量更是达到48种，其次是云南，达到39种，种类数量的优势如此之大，所以在中国观鸟活动的推荐目的地中，西南山地是当之无愧的首选目的地。

这里我将根据多年的观鸟经验，整理出西南山地的五条观鸟拍鸟线路，涵盖四川、青海、西藏和云南的大部分地区，包揽所有在西南山地区域分布的60种严格意义上的中国特有种，并加上另外30种濒危、极危和稀有的不属于特有种的鸟种，来和读者分享。

线路一：西南山地之四川中、西、北部

四川位于中国大陆青藏高原和长江中下游平原的过渡带，高低悬殊，西高东低的特点明显。西部、北部为高原、山地，海拔多在3000米以上；中部、东部为盆地、丘陵，海拔多在500米至2000米之间；境内高山、高原、森林、平原、湿地和荒漠等生境多样，60%以上的中国特有鸟种分布在四川。

成都肯定是一个理想的出发集结地，不仅仅有"来了就不想离开"的美誉，市区、市郊的各大城市绿地公园，都有很好的鸟类观赏机会，根据最新统计，成都市的鸟种记录已经达到480多种。百花潭公园和浣花溪公园值得你花费整整一个上午，植物园和熊猫基地都完全可以安排一个整天，还有东边青龙湖大面积的湿地，往往会有意外的收获。如果是迁徙季节，你来到成都，一定要去川大望江校区的"天使林"守望，说不定平日里很难在野外繁殖地目睹的珍稀夏候鸟，诸如金胸歌鸲（Calliope pectardens）和黑喉歌鸲（Calliope obscura）等能在这里和你不期而遇，就算是棕头歌鸲（Larvivora ruficeps），也不是完全没有可能邂逅。距离成都很近的北边的鸭子河流域，冬季可是很多雁鸭类的越冬之地，有着"鸟中大熊猫"之称的青头潜鸭（Aythya baeri），在鸭子河几乎每年都有稳定的记录；周围河谷的清澈山涧之中，不难寻到中华秋沙鸭（Mergus squamatus）的踪迹；还有冬季垂直迁徙到低海拔地带的鹮嘴鹬（Ibidorhyncha struthersii）行走在河边沙砾之上。

看完成都的鸟，品尝完成都美食，下一步就是向西南穿过成都平原进入雅安市荥经县的龙苍沟国家森林公

---

① 郑光美：《中国鸟类分类与分布名录（第二版）》，科学出版社，2011年。

园,这是个最近几年才被国内、国际观鸟人士发现的观鸟胜地,位于邛崃山脉西南方向的大小相岭保护区范围之内。从公园入口海拔1200米左右的农田、常绿阔叶林区域到景区公路尽头的大相岭保护区的大熊猫野化放归基地,短短的15公里,海拔上升到2500多米,生境变换为以针叶林和竹丛为主。从景区入口到距离入口6公里处的自然宣教中心的区域,依次可以寻觅到灰胸竹鸡(*Bambusicola thoracicus*)、四川短翅蝗莺(*Locustella chengi*)、黄腹山雀(*Pardaliparus venustulus*)、棕噪鹛(*Pterorhinus berthemyi*)、白腹锦鸡(*Chrysolophus amherstiae*)、峨眉柳莺(*Phylloscopus emeiensis*)、金额雀鹛(*Schoeniparus variegaticeps*)、褐顶雀鹛(*Schoeniparus brunneus*)等特有珍稀鸟种。一早来到森林公园大门附近,灰胸竹鸡(*Bambusicola thoracicus*)在农田边的灌丛中此起彼伏地鸣叫,四川短翅蝗莺(*Locustella chengi*)在草丛中不断发出特有的拉锯一样的声音,黄腹山雀(*Pardaliparus venustulus*)和峨眉柳莺(*Phylloscopus emeiensis*)就在自然宣教中心旁的树端不断跳跃,地面上白腹锦鸡(*Chrysolophus amherstiae*)沿着土路在觅食,棕噪鹛(*Pterorhinus berthemyi*)成群聒噪地从这棵树飞向另外一棵树,金额雀鹛(*Schoeniparus variegaticeps*)就在溪水边阴湿的竹林下一闪而过,只留给你一个小黑点的印象,让你全然不知道为啥叫"金额",好在还有褐顶雀鹛(*Schoeniparus brunneus*)在竹丛根部钻进钻出。而后继续沿着生态土路上行到距离入口10公里左右的区域,在松树的顶端,酒红朱雀(*Carpodacus vinaceus*)就站在那里,灰胸薮鹛(*Liocichla omeiensis*)在松林和竹林混合的区域悠扬地歌唱,红腹角雉(*Tragopan temminckii*)很有可能就在土路旁边的某个小土墩上面,当然背后肯定有浓密的竹林,三趾鸦雀(*Cholornis paradoxus*)、红翅噪鹛(*Trochalopteron formosum*)、黄额鸦雀(*Suthora fulvifrons*)和灰头雀鹛(*Fulvetta cinereiceps*)在同一片竹林里面逡巡,你只需要安静地等候在那里,看它们几乎从你面前由土路的这一侧飞进另外一侧的竹林,此时你可不要忽略,就在不远处的一棵小树上,有两只红腹山雀(*Poecile davidi*)正安静地寻找着虫子。由此到公路尽头的最后5公里区域,是这个观鸟热点地区最为耀眼的两种特有鸟——暗色鸦雀(*Sinosuthora zappeyi*)和灰头斑翅鹛(*Actinodura souliei*)的分布区域,暗色鸦雀总是忙碌地活动在小树和竹林交汇处,如果旁边有树干上布满青苔的大松树,多仔细看看,这里可是灰头斑翅鹛最喜欢的觅食之处。树林和竹林密集并有杜鹃树的沟谷是中华鹪鹛(*Pnoepyga mutica*)的地盘,在土路尽头的原始森林之中,不时可见四川旋木雀(*Certhia tianquanensis*)忙碌地顺着树干爬上爬下,白领凤鹛(*Patayuhina diademata*)总是成对地在小树上观察着你。

由龙苍沟国家森林公园往西北方向不到2小时车程就是二郎山喇叭河景区,这里是华西雨屏带的核心区域，是一种比较罕见的气候地理单元，生物多样性异常丰富，与印度的大吉岭地区并称为"地球两大物种基因库"。景区里面吃住行的条件很是成熟便利，从核桃坪延伸而上的生态步道，灌丛、竹林交替变换，这里更容易见到白腹锦鸡（*Chrysolophus amherstiae*）、红腹角雉（*Tragopan temminckii*）、四川旋木雀（*Certhia tianquanensis*）和蓝鹀（*Emberiza siemsseni*），还有平时难睹芳容的金胸歌鸲（*Calliope pectardens*），不难从传来婉转歌声的地方找到它，并惊艳于那胸前的一抹红；越来越难见的黄额鸦雀（*Suthora fulvifrons*），在步道的尽头区域还相对容易见到。距离喇叭河景区不远的二郎山翻山老路，行至此处就会看到宝兴鹛雀（*Moupinia poecilotis*）在低海拔灌丛中穿梭，宝兴歌鸫（*Turdus mupinensis*）在中海拔针叶林的顶端歌唱，淡腹点翅朱雀（*Carpodacus verreauxii*）在尖利、高亢、脆生的鸣叫中，飞行于高海拔区域的矮小树丛之间。

现在我们就该转道北上了，翻越喇叭河景区北边海拔高达4114米的夹金山垭口，上山途中的悬崖一带是寻找红翅旋壁雀（*Tichodroma muraria*）的理想区域。山顶的高海拔地区，藏雪鸡（*Tetraogallus tibetanus*）肯定会在某个流石滩驻足，如果你运气够好，再加上努力寻找，这里就是一个高概率可以目睹绿尾虹雉（*Lophophorus lhuysii*）的地方。下山的途中，进入针叶林，安静地走在松针绵绵的地面上，肯定会惊飞几只栗背岩鹨（*Prunella immaculata*），但它们又会降落在不远处的地面上，抑或是看见几只红喉雉鹑（*Tetraophasis obscurus*）笨拙地跑进密林深处。

而后我们将抵达汶川的卧龙，这里是以大熊猫为最主要保护物种的保护区，坐落在岷山山脉的腹地，从海拔1200米的皮条河流域一直到达海拔4480米的巴朗山垭口，区区100公里之间，海拔落差超过3000米，其间交替分布着常绿阔叶林、常绿落叶阔叶混交林、针叶阔叶混交林、寒温性针叶林、耐寒灌丛和高山草甸以及高山流石滩稀疏植被带6种植被，垂直分布的不同植被生境以及便利的交通条件，使得这里成为观鸟爱好者的天堂，以卧龙镇所在的沙湾为基地，这里值得你停留2—3天。

低海拔区域的喇嘛庙附近，要见到红腹锦鸡（*Chrysolophus pictus*）、灰胸竹鸡（*Bambusicola thoracicus*）、蓝鹀（*Emberiza siemsseni*）甚是容易，早上下山觅食的红腹角雉（*Tragopan temminckii*）也有很大的几率遇到；就连不常出现的棕头雀鹛（*Fulvetta ruficapilla*），说不定也正在寺院旁边的灌丛中成小群移动。而在寺院外河对面的南岸山坡上，分布着低矮的灌丛和浓密的竹林，这里是斑背噪鹛（*Ianthocincla lunulata*）、金胸歌鸲（*Calliope pectardens*）、红腹锦鸡（*Chrysolophus pictus*）的乐园。更让人激动的是，根据2015年春季的记录，这里有种群数量很小的黑喉歌鸲（*Calliope obscura*）繁殖，只是生性警觉的它们总是躲在浓密竹林的深处，观鸟人也只能闻声兴叹。

中高海拔区域的贝母坪、花岩子隧道口一带的针叶林、针阔混交林中分布着生性极为警觉的斑尾榛鸡（*Tetrastes sewerzowi*），它是野外观赏难度极高的一种雉类，需要在静谧的清晨，首先听到尖利而断断续续的独特叫声，而后去寻。沿着公路慢慢前行，你会发现红腹角雉（*Tragopan temminckii*）在公路沿线的草地上觅食，暗红色带斑点的体羽在清晨的雾霭之中愈发靓丽，在发现了你的存在时，它才恍然大悟般转身步入灌丛，留给你无尽遐想；红喉雉鹑（*Tetraophasis obscurus*）有时会快步横穿公路，消失在密林之中，或者站立在公路旁的小土堆之上，这会儿你就需要注意土堆旁边了，应该会有其他的红喉雉鹑在附近觅食。在花岩子隧道口附近山上的林缘草地上，伴随着旋律多变而婉转悦耳的叫声，不难在阳光下观赏到绿尾虹雉（*Lophophorus lhuysii*）特有的带金属光泽的羽毛在熠熠发光，它时而低头觅食，时而站立不动，时而滑翔而过，留给你一个美妙的剪影；白马鸡（*Crossoptilon crossoptilon*）成对或者成群在草地上活动，显得格外醒目，嘹亮的叫声让你不会错过它们那优雅的姿态；小羊般咩咩叫的白眉朱雀（*Carpodacus dubius*）是你肯定不会错过的燕雀科鸟类；硕大的大噪鹛（*Ianthocincla maximus*）鸟如其名，大而吵闹，很远就能听见它们的叫声，并且可以近距离看它们在草地上觅食，或穿梭于灌丛中；隧道外侧的崖壁周围是淡腹点翅朱雀（*Carpodacus verreauxii*）、中华雀鹛（*Fulvetta striaticollis*）、华西柳莺（*Phylloscopus occisinensis*）、棕背黑头鸫（*Turdus kessleri*）的出没之地；过隧道之后的山坡，如果是在一个雾蒙蒙的凌晨，说不定你可以听见林沙锥（*Gallinago nemoricola*）很有节奏的"哒-哒-哒"的叫声，并偶见其行走在湿漉漉的草地花丛之中。

翻过巴朗山，我们继续北上抵达距离马尔康40公里的梦笔山森林公园，从海拔高达4480米的垭口蜿蜒而下的15公里的区域是这片森林中最主要的观鸟地点。清晨来到这里，高海拔林缘区域有稀疏灌丛的草地上，中华朱雀（*Carpodacus davidianus*）、曙红朱雀（*Carpodacus waltoni*）在这里觅食，如果你靠近，它们会就近飞上灌丛，给你更好的观察角度；沿路而下，你或可以听见红喉雉鹑（*Tetraophasis obscurus*）此起彼伏地在林下鸣叫，或看见它们陆续穿过公路；针叶林中传来细丝般的声音，那一定是凤头雀莺（*Leptopoecile elegans*），它总是忙碌地从一簇针叶钻进另外一簇，不愿意给你一个色彩绚丽的全身像；忽然，你可能会听见"嘎"的一声长鸣，一个黑色倩影翩然滑翔而至，对，那就是黑头噪鸦（*Perisoreus internigrans*），仅见于青藏高原东部的特有留鸟；继续沿路而下，公路两侧的灌丛之中不难寻找到斑翅朱雀（*Carpodacus trifasciatus*）的身影，它也许正站在灌丛上咀嚼着嫩叶。

从马尔康开始，慢慢沿着河谷上行，进入青藏高原的东南低矮部分，抵达中途的红原的时候，就已经置身于闻名遐迩的青藏高原了，鸟的种类相对于低矮谷地已经有很大的变化，一路欣赏着黑颈鹤（*Grus nigricollis*）的欢歌曼舞，追寻着路边灌丛中的白眉山雀（*Poecile superciliosus*）这个高原特有种，我们将抵达若尔盖这个国内最大的高原湿地中心。

东出若尔盖县城不到30公里，沿着一个陡降的山谷前行，进入一片原始针叶林，这里可是观鸟者的另一个天堂，名字叫"巴西森林"。你会发现蓝马鸡（*Crossoptilon auritum*）优雅的身影就在阳光下森林边的草地上；清晨四川林鸮（*Strix davidi*）就在用浑厚的叫声宣示着领地不可侵犯，斑尾榛鸡（*Tetrastes sewerzowi*）在针叶林中扑腾着，你循声而去，总会见到它奔跑着消失在树林深处；黑额山噪鹛（*Ianthocincla sukatschewi*）不难在森林和灌丛交汇处看见；在枝头不停歌唱的，总是四川褐头山雀（*Poecile weigoldicus*）；白脸䴓（*Sitta przewalskii*）总是执着地守着一棵干枯的老树，上下寻觅爬行，时而翘首远望；山噪鹛（*Pterorhinus davidi*）就在村庄旁的牛圈栏杆上跳上跳下；当你仰望松树上端的黑头噪鸦（*Perisoreus internigrans*）的时候，还需要注意树下的灌丛，那里面可能有一群白眶鸦雀（*Sinosuthora conspicillata*）正在向你靠近。

西出若尔盖到花湖的这60公里的路程完全是高原草地和湿地的景观，众多的地山雀（*Pseudopodoces humilis*）在地上一蹦一站，或者飞上沿路的水泥桩，站在那里打量着你；棕颈雪雀（*Pyrgilauda ruficollis*）蹦蹦跳跳，全然不顾你的存在；白腰雪雀（*Onychostruthus taczanowskii*）会从废弃的高原鼠兔（*Ochotona curzoniae*）的巢穴里探出一个脑袋；偶尔低空飞过的猎隼（*Falco cherrug*）和草地上停驻的草原雕（*Aquila nipalensis*），才会让棕颈雪雀一哄而散，白腰雪雀俯身潜入高原鼠兔废弃的巢穴。

南下若尔盖50公里的地方是连绵不断的草地缓坡，生长着一些稀疏低矮的灌丛，在很久之前就有繁殖记录的朱鹀（*Urocynchramus pylzowi*），几年前终于在这个稳定的繁殖地点被重新记录；而且这里也是藏雪雀（*Montifringilla henrici*）分布地域的最南端。

从若尔盖往东，是中国第一个以保护自然风景为主要目的的自然保护区——九寨沟自然保护区，这里曾经是稀有的棕头歌鸲（*Larvivora ruficeps*）的稳定繁殖地点，但是不知道什么原因，从2016年开始就没有再在这里记录到这个稀有鸟种，但是仍然值得确信的是，在附近的某个暂时还不为人知的合适的生境，它们一如既往地在空旷的森林中怡然歌唱。九寨沟附近的弓杠岭森林是另外一个值得停留的地方，无论是四川林鸮（*Strix davidi*）、斑尾榛鸡（*Tetrastes sewerzowi*）、蓝马鸡（*Crossoptilon auritum*）以及黑头噪鸦（*Perisoreus internigrans*），抑或是斑翅朱雀（*Carpodacus trifasciatus*）、红腹山雀（*Poecile davidi*）、银脸长尾山雀（*Aegithalos fuliginosus*）、黑额山噪鹛（*Ianthocincla sukatschewi*），都有很大的概率可以目睹。

接着我们继续东进抵达唐家河自然保护区，在这个以保护大熊猫、金丝猴和扭角羚及其栖息地为主要目的的保护区，分布着极为稀少的特有的灰冠鸦雀（*Sinosuthora przewalskii*），但是由于有可能目睹到的地方在保护区的核心区域内，按照法律规定是不能对公众旅游开放的，所以需要提前和保护区联系。但是除此之外的区域，比如景区公路沿线和阴平古道的生态徒步栈道，都是很好的观鸟地点，特别是阴平古道，沿着幽静的石梯拾级而上，红腹锦鸡（*Chrysolophus pictus*）、红腹角雉（*Tragopan temminckii*）不时跨过栈道，甚至于就在离你不远处继续觅食，全然不介意你的出现；再往上面一点，斑背噪鹛（*Ianthocincla lunulatus*）、蓝鹀（*Emberiza siemsseni*）和白眶鸦雀（*Sinosuthora conspicillata*）在林中跳跃。当你置身于古栈道的顶端的时候，理论上，这里是灰冠鸦雀（*Sinosuthora przewalskii*）的栖息地，有待你的发现。在景区大门外的村庄、农田这些低海拔区域，则分布着画眉（*Garrulax canorus*）和灰胸竹鸡（*Bambusicola thoracicus*）。

最后这条四川中、西、北部区域的特有鸟种观察线路以从唐家河自然保护区经由G5高速公路返回成都结束，到了成都，你就又可以再次体会这个"来了就不想离开"的城市的各种魅力了。

线路二：西南山地之四川南部

依旧还是将成都作为这条线路的集结地，我们将一路向南进入凉山山系区域，遍览川南特有鸟种。

首先我们将直接抵达宜宾屏山的老君山这个以四川山鹧鸪（*Arborophila rufipectus*）为保护对象的国家级自然保护区，保护区地处全球生物多样性热点地区之一的凉山山系，老君山主峰海拔2008.7米，是川南第一高峰，山顶常年烟云霭霭，雾锁葱茏，高山耸峙，深谷回响，悬泉瀑布飞漱其间，是四川亚热带常绿阔叶林原始林保存最完好的片区之一，也是川南唯一的亚热带原始林区。抵达前，你将轻松地在公路旁、树林边收获与领雀嘴鹎（*Spizixos semitorques*）的邂逅；而后我们将直接前往半山的二燕坪一带，如果运气不错，就可以一睹四川山鹧鸪（*Arborophila rufipectus*）的风采，但通常还是得沿着土路和石阶跋涉上到山顶一带，在静谧的清晨，缓步细听竹林下窸窸窣窣的声音，那可能就是一只警觉的四川山鹧鸪正在觅食，而你所需要做的是安静地蹲下来，透过竹林的空隙，期望它步入你的视线。老君山还是灰头斑翅鹛（*Actinodura souliei*）、金额雀鹛（*Schoeniparus variegaticeps*）、灰胸薮鹛（*Liocichla omeiensis*）和棕头雀鹛（*Fulvetta ruficapilla*）等珍稀鸟种野外目睹几率很大的地方，在下山的过程中，它们将不时出现在你的视野中。

离开老君山之后，我们将深入凉山山系抵达马边县菝坝镇，根据学者近几年的研究发现和监测，这里是珍稀的夏候鸟鹊鹂（*Oriolus mellianus*）夏季稳定的繁殖地点，沿着乡间的安静土路，鹊鹂嬉戏于亚热带常绿阔叶林之中。

告别了鹊鹂，我们继续穿越凉山山系直抵位于横断山脉最南端的木里县，这一范围内整体地势西高南低，四条南北走向的山脉相间排列，河流深切，岭谷相对高差很大。在这里的干热河谷一带，分布着稀有的大紫胸鹦鹉（*Psittacula derbiana*），由于它们分布范围不广，加上人为盗捕的影响，其野外数量逐渐下降的趋势很是明显。还有中国特有鸟种白点噪鹛（*Ianthocincla bieti*），尽管模式产地是四川中部的宝兴，但是现在也只在四川西南部的木里、云南西北部的丽江一片狭小区域内有分布记录，从木里县城往西北木里大寺方向大约40公里左右有一片浓密的竹林，由于很少有外界的干扰，在这里目睹和拍摄到白点噪鹛的几率还是很大的。

从木里的大山出来后，可以就近到西昌结束这次的探访，或者南下云南展开另外一段追寻精灵的旅途。

线路三：西南山地之云南中西部

云南，这个人类文明重要的发祥之地，境内山脉逶迤，江河蜿蜒，森林茂盛，气候类型多样，为各种动物的生存、繁衍提供了得天独厚的自然环境，动物种类在全国居于首位。

云南整体地势由东南向西北抬升，气候、植被等环境因素也相应发生变化，从热带雨林渐次向亚热带常绿阔叶林、温带针阔混交林、寒温性常绿阔叶林、高寒草原草甸景观类型过渡。根据最新的数据显示，云南记录鸟类种数接近850种，其中以常绿阔叶林生境类型为最多，占比一半以上。

笔者于2005年第一次前往云南观鸟，后来又多次前往，不管是多绚丽热带鸟种的西双版纳地区，抑或是多高原山地鸟种的云南西北地区，还是云南西南一片的鹛类天堂，笔者都曾涉足。但是其中最推荐的仍要数昆明、楚雄、丽江、大理、泸水、保山、盈江和瑞丽一线。

以昆明作为整条线路的起点，而后前往楚雄的紫溪山森林公园，交通很是便利。景区内峰峦起伏，森林茂密，古木参天。清晨漫步在林荫小道上，滇䴓（*Sitta yunnanensis*）成对站在松树上，巨䴓（*Sitta magna*）响亮独特的叫声让你不会错过它那浓黑的眉毛，还有栗背岩鹨（*Prunella immaculata*），就在某座小桥下的草地上觅食，半天或是一整天都在。不过除非为了得到很满意的照片，否则还是建议尽快前往下一站。

抵达丽江后，充分利用当地特有的腊排骨积蓄的力量，一早前往永胜方向距离市区40多公里的万亩杜鹃园，这里可是国内最靠谱的白点噪鹛（*Ianthocincla bieti*）的观鸟点，相信你不会错过和它的邂逅。园内开阔处旁边的低矮灌丛夹杂松树林就是白点噪鹛典型的生境，甚至于公路旁的类似环境，也可能见到它从里面慢慢跳出来。在这里我们还将看见棕头雀鹛（*Fulvetta ruficapilla*）在灌丛中穿梭，宝兴鹛雀（*Moupinia poecilotis*）在灌木的枝头上随风颤颤。在丽江市区的黑龙潭公园，也可以见到棕头雀鹛（*Fulvetta ruficapilla*）；如果前往不远的"长江第一湾"的石鼓镇，途中的松林里也容易见到滇䴓（*Sitta yunnanensis*）。

离开丽江，我们直奔大理，苍山无为寺后面的羊肠小道上的白腹锦鸡（*Chrysolophus amherstiae*）、宝兴鹛雀（*Moupinia poecilotis*），以及晚上洱海边的风花雪月啤酒，都会使你愿意停留1—2天。

继续沿西南方向行进，将进入高黎贡山山脉，它是横断山脉中最西部的山脉，北连青藏高原，南接中南半岛，山高坡陡切割深，垂直高差达4000米以上，形成极为壮观的垂直自然景观和立体气候。高黎贡山山脉跨越五个纬度，是地球上迄今唯一保存有大片由湿润热带森林过渡到温带森林的过渡带的地区，是世界上极其珍贵也极其稀有的生物多样性十分突出的地区。东西坡海拔1600—2800米的区域是中国最引人瞩目的原始阔叶林区，400余种鸟类世代生活在这里，其中不乏很多中国特有和珍稀鸟种。

抵达保山之前建议先到泸水，翻越片马丫口，抵达边境小镇，铜锅牛肉自然会洗净你奔波的疲惫。而后建议用2—3个整天返回片马丫口两侧逡巡，在这里不仅有可能偶遇白尾梢虹雉（*Lophophorus sclateri*），而且火尾绿鹛（*Myzornis pyrrhoura*）也不时往返于公路两侧的低矮灌丛之中，在相对低海拔的竹林区域，黄额鸦雀（*Suthora fulvifrons*）更是较为普遍地分布着。

过了片马丫口之后，下一个目的地是保山的百花岭。在宁静的南斋公房古道，主要是旧街子和大炉厂一带，建议使用上下来回不停慢步移动这种最没有技术含量的观鸟拍鸟方式，数不尽的鹛类和其他一些稀有鸟种，会让你乐此不疲，流连忘返。

下一站应该是盈江了，有两个地点一定要去，一个是经典的那邦小镇，包含榕树王和铜壁关，还有就是新近很火的红崩河区域。而盈江之后就是最后一站的瑞丽了，不管是莫里的热带雨林，还是南京里已经被开发后残存的区域，以及海关旁边的姐告大桥一带，尽管这些区域没有多少中国特有鸟类的分布，但是有许多稀有难见以及可能是新近记录的鸟类，可能会给你带来不少惊喜。

到这里应该可以给鸟友们的旅途画上一个较为圆满的句号了，而后就可以踏上回家的路，因为除了钟爱的野外精灵之外，还有更为关爱你的家人在等着你。

线路四：西南山地之青海

青藏高原平均海拔4500米以上，从西北至东南，海拔由高到低，植被从高山寒漠逐步演替为荒漠草原。而东北部的青海祁连山脉一直向南到喜马拉雅山脉与横断山脉，多切割山地，高山带以下多为草原，东部和东南部分布有高山针叶林、灌丛和草甸，构成了极其多样的生境，造就了其特殊的鸟类区系以及丰富的特有鸟种。再加上藏族人民独特的风俗习惯以及这一地区浓郁的藏传佛教氛围，使得青藏高原的鸟类与人类的距离更加接近，在高原观察鸟类便有了得天独厚的优势。

以青海省的省会西宁为起点，这里海拔2261米，位于湟水谷地。毗邻城市北边的北山以及附近的大通鹞子沟国家森林公园和互助北山国家森林公园区域内，不难观测到沙色朱雀（*Carpodacus stoliczkae*）在干旱的斜坡公路边活动；斑翅山鹑（*Perdix dauurica*）鸣叫着在草丛中跑远；甘肃柳莺（*Phylloscopus kansuensis*）这一特有种在枝头上老是不停地飞动；黑头䴓（*Sitta villosa*）、白脸䴓（*Sitta przewalskii*）还有绚丽的凤头雀莺（*Leptopoecile elegans*）总是在针叶林上急急忙忙地钻来钻去；斑尾榛鸡（*Tetrastes sewerzowi*）生性警觉，但是清晨走进互助北山的原始针叶林，还是不难见到，甚至一次看见几只雏鸟，而斑尾榛鸡妈妈肯定就在旁边，并且还尽力吸引着你的注意力；山噪鹛（*Pterorhinus davidi*）在谷口的低矮灌丛之中蹿来蹿去；互助北山的十二盘坡是观察蓝马鸡（*Crossoptilon auritum*）的绝佳地点，每个早晨，它总会在阳光下，从森林的边缘踱步出来，穿越一片草地，慢步进入另外一片灌丛，还不时发出粗狂的警觉鸣叫。

而后从西宁前往青海湖，在青海湖东侧的沼泽地带，可以观测到高原神鸟黑颈鹤（*Grus nigricollis*），这唯一在高原繁殖的鹤类优雅地散着步，时而引吭高歌，时而振翅欲飞；新近从崖沙燕亚种提升为新种的淡色崖沙燕（*Riparia diluta*）总是飞得太高太快，观鸟者得睁大眼睛仔细辨认。

青海湖西侧的黑马河可以作为短时的观鸟据点，特别是西边的橡皮山河谷一带，可以观察到独特的朱鹀（*Urocynchramus pylzowi*），它是朱鹀科（*Urocynchramidae*）下唯一的一个种类，许多国际国内的观鸟爱好者千里迢迢来到这里，就是为了一睹它阳光下绚烂的红色，看它在空中缓慢飞行、不停振翼；由黄腹柳莺亚种提升为新种的华西柳莺（*Phylloscopus occisinensis*）在此地倒是很容易看见；白眉山雀（*Poecile superciliosus*）这一高原特有种在这里是区域性常见鸟种；地山雀（*Pseudopodoces humilis*）沿着废弃的公路两侧不停跳跃前进，不时跃上一座土堆，回头打量着你；公路边的土崖，不时有红翅旋壁雀（*Tichodroma muraria*）光临。

从黑马河向西不远是茶卡盐湖区域，位于柴达木盆地的东部，在荒漠草丛地带，不难观察到特有的黑尾地鸦（*Podoces hendersoni*），它总是快步前驱，奔跑在干涸的地面上，偶尔会飞行一段较短的距离，停留在枝头或者小栏杆上，再回头望你，刻意与你保持一个安全距离；在茶卡盐湖周围的山谷，特别是以北27公里左右的山谷中，凌晨总能听见大石鸡（*Alectoris magna*）这一特有鸟种的高亢鸣叫，并见它们傲立于石头的顶端；还有贺兰山红尾鸲（*Phoenicurus alaschanicus*），清晨的淡雾中，总能目睹它神奇的蓝和红。

从茶卡盐湖区域折返回214国道，翻越巴颜喀拉山口，抵达鄂拉山口，这里海拔超过4500米，昔日为唐蕃古道要隘，植被以高山草甸为主。

凌晨来到这里，即使是盛夏的早上，气温依然可能在零度以下，而且寒风阵阵，需要沿着山脊艰难地慢慢往上攀爬到山口，在第一缕阳光下，你会被藏雀（*Carpodacus roborowskii*）的深红所吸引，小精灵就在不多的高原草甸上，蹦跳着寻找新鲜的嫩叶；从这里还得继续向上到山顶（海拔约5000米），抵达一片开阔广袤之地，这里就是青藏高原特有的西藏毛腿沙鸡（*Syrrhaptes tibetanus*）的栖息地，只要耐心地找寻和等待，基本都能看见沙鸡飞翔起来，然后降落在不远处，如果需要仔细观察西藏毛腿沙鸡的一应特征，就需要慢慢靠近、靠近、靠近，不惊扰它，在你的望远镜之中，你会看见皮黄色的羽毛之上居然有神奇的纹路。这时的你应该已经是心满意足，该坐下来静静体会一下，聆听自然的声音。但是你多半会发现，白腰雪雀（*Onychostruthus taczanowskii*）、棕颈雪雀（*Pyrgilauda ruficollis*）和藏雪雀（*Montifringilla henrici*）就在你周围的高原草甸上跳行，特别是白腰雪雀，它们会利用废弃的高原鼠兔的巢穴来哺育自己的下一代，你总会见到它们漂亮的白肚皮，忙碌地在巢穴口进进出出。

从鄂拉山口继续沿214国道南下，过称多县珍秦镇之后不久，就告别高原地区，下行到青藏高原东南部的河谷地带，这里海拔逐渐走低，植被多为森林和灌丛。经过玉树之后，在河谷之中，你会发现长嘴的鹮嘴鹬（*Ibidorhyncha struthersii*），头一颤一颤地在河边的草甸上努力找寻着什么；河谷旁边的高大岩石下端的草地上，拟大朱雀（*Carpodacus rubicilloides*）在阳光下极力地炫耀着那刺眼的红色羽毛。

下一站是囊谦，它西接青藏高原，东临横断山脉，澜沧江贯穿全境，地理和生境资源极其丰富，有雪山、湖泊、草地和森林。在距离囊谦县城40公里左右的坎达山口海拔较高的地方，我们将看到高原特有的藏鹀（*Emberiza koslowi*）在清晨的阳光中纵情歌唱；从坎达山口下到低海拔的森林地带，总能看见黄喉雉鹑（*Tetraophasis szechenyii*）在林间的凸出草地上聒噪地持续鸣叫，宣示着主权；棕草鹛（*Pterorhinus koslowi*）——这全身棕色的高原特有噪鹛，不停穿梭于林下或者跳跃于草地上。

距离囊谦县城60公里左右的白扎林场是另外一处观鸟胜地，坐落在宁静的沟谷中，有一座古老的藏传佛教寺院——尕尔寺，白眉朱雀（*Carpodacus dubius*）、凤头雀莺（*Leptopoecile elegans*）、橙翅噪鹛（*Trochalopteron elliotii*）欢快的叫声，在那里衬托着这个古老寺院的静谧和森林的宁静。

线路五：西南山地之西藏

拉萨作为具有高原和民族特色的国际旅游城市，以风光秀丽、历史悠久、风俗民情独特、宗教色彩浓厚而闻名于世，这里全年多晴朗天气，降雨稀少，冬无严寒，夏无酷暑，气候宜人，生物资源极其丰富。

西藏特有珍稀鸟类观赏线路以拉萨为起点，包括周边的河谷、湿地地区，然后沿318国道向东，进入藏东大峡谷地区的森林和湖泊，丰富的生境资源造就了拉萨—林芝的高原鸟类野外观察的黄金线路。环线返回拉萨或者在林芝结束，建议时间都是一周左右。

在拉萨附近往西不远的一处河谷，有个叫雄色寺的尼姑寺，凌晨前从拉萨赶到这里，便可见到藏马鸡（Crossoptilon harmani）在山谷的突兀岩石上伫立高歌，或者沿着转经道追随着你的脚步；大草鹛（Pterorhinus waddelli）在山坡的灌丛中不停穿梭，不时会来到开阔的空地，向你展示它嘹亮的歌喉；灰腹噪鹛（Trochalopteron henrici）带着它那橙红的臀，还有那让人过目不忘的眼下的一道明显的白斑，此时就在你的周围蹦跳；从乌鸫亚种独立出来的藏乌鸫（Turdus maximus），到处都是它们的身影；这里的山谷里，还有花彩雀莺（Leptopoecile sophiae）在灌丛中不停穿梭，红红的拟大朱雀（Carpodacus rubicilloides）在枝头上跳跃，山谷下的农田里，成群的高原山鹑（Perdix hodgsoniae）守候在田间空地。

从拉萨沿318国道往东，经过海拔5013米的米拉山口，山口附近的草地上，白腰雪雀（Onychostruthus taczanowskii）走走停停，地山雀（Pseudopodoces humilis）驻足回望，棕颈雪雀（Pyrgilauda ruficollis）更是不难一见。

抵达工布江达县的巴松措高原湖泊区域，这里不仅仅有人间自然美景，当你漫步湖边，或者徒步进入林区，往措宗寺方向走，到处可以不经意地看见独特的高原鸟类，黑颈鹤（Grus nigricollis）在这里悠闲地繁殖。

继续沿318国道东进到林芝，这里平均海拔3000米左右，坐落于尼洋河和雅鲁藏布江交汇处，温暖的印度洋暖流沿雅鲁藏布江逆行而上，通过重重障碍，贯穿喜马拉雅山东段，造就了这个被称为"西藏江南"的世外桃源。

继续往东不到80公里，就抵达海拔高达4728米的色季拉山口，短距离、高落差造就了这一带丰富的生境资源，观看成群的大紫胸鹦鹉（Psittacula derbiana）是很多观鸟者来到这里的重要目的。

从色季拉山口沿318国道一直到波密，路程200公里左右，沿路几乎每个地点都可以停下来，就在公路的两侧，任意选择一个山谷，步入其中，就可以充分享受这项审美体验与科学探索相统一的新兴的生态活动——观鸟。

拉萨—林芝黄金线路结束后，可以选择从林芝乘机离开或者返回拉萨。

附录二 索引

中文名索引

A

暗色鸦雀 / 100

B

白点噪鹛 / 076
白腹锦鸡 / 034
白眶鸦雀 / 098
白脸鸻 / 110
白领凤鹛 / 106
白马鸡 / 028
白眉朱雀 / 138
白尾梢虹雉 / 022
斑背噪鹛 / 078
斑翅朱雀 / 136
斑尾榛鸡 / 010
宝兴歌鸫 / 118
宝兴鹛雀 / 088

C

橙翅噪鹛 / 066

D

大噪鹛 / 074
大紫胸鹦鹉 / 042
淡腹点翅朱雀 / 134
滇䴙 / 108

E

峨眉柳莺 / 056

H

黑额山噪鹛 / 072
黑喉歌鸲 / 124
黑颈鹤 / 038
黑头噪鸦 / 046
红翅噪鹛 / 062
红腹角雉 / 020
红腹锦鸡 / 032
红腹山雀 / 048
红喉雉鹑 / 012
画眉 / 070
黄额鸦雀 / 104
黄喉雉鹑 / 014
灰冠鸦雀 / 102
灰头斑翅鹛 / 064
灰头雀鹛 / 094

灰胸薮鹛 / 068
灰胸竹鸡 / 018
火尾绿鹛 / 084

J

金额雀鹛 / 060
金胸歌鸲 / 126
巨䴙 / 112

L

蓝马鸡 / 030
蓝鹀 / 140
栗背岩鹨 / 128
林沙锥 / 040
领雀嘴鹎 / 052
绿尾虹雉 / 026

N

拟大朱雀 / 130

Q

青头潜鸭 / 004
鹊鹂 / 044

S

三趾鸦雀 / 096
曙红朱雀 / 132
四川短翅蝗莺 / 058
四川褐头山雀 / 050
四川山鹧鸪 / 016
四川旋木雀 / 114

Y

银脸长尾山雀 / 054

Z

中华秋沙鸭 / 008
中华雀鹛 / 090
棕背黑头鸫 / 116
棕草鹛 / 080
棕头歌鸲 / 120
棕头雀鹛 / 092
棕噪鹛 / 082

英文名索引

B

Baer's Pochard / 004

Barred Laughingthrush / 078

Black-necked Crane / 038

Blackthroat / 124

Blue Eared Pheasant / 030

Buff-throated Monal Partridge / 014

Buffy Laughingthrush / 082

C

Chestnut-throated Monal Partridge / 012

Chinese Bamboo Partridge / 018

Chinese Fulvetta / 090

Chinese Grouse / 010

Chinese Monal / 026

Chinese Thrush / 118

Chinese White-browed Rosefinch / 138

Collared Finchbill / 052

D

Derbyan Parakeet / 042

E

Elliot's Laughingthrush / 066

Emei Leaf Warbler / 056

Emei Shan Liocichla / 068

F

Fire-tailed Myzornis / 084

Firethroat / 126

Fulvous Parrotbill / 104

G

Giant Laughingthrush / 074

Giant Nuthatch / 112

Golden Pheasant / 032

Golden-fronted Fulvetta / 060

Grey-hooded Fulvetta / 094

Grey-hooded Parrotbill / 100

H

Hwamei / 070

K

Kessler's Thrush / 116

L

Lady Amherst's Pheasant / 034

M

Maroon-backed Accentor / 128

P

Pink-rumped Rosefinch / 132

Przevalski's Nuthatch / 110

R

Red-winged Laughingthrush / 062

Rufous-headed Robin / 120

Rufous-tailed Babbler / 088

Rusty-breasted Tit / 048

Rusty-throated Parrotbill / 102

S

Scaly-sided Merganser / 008

Sclater's Monal / 022

Sharpe's Rosefinch / 134

Sichuan Bush Warbler / 058

Sichuan Jay / 046

Sichuan Partridge / 016

Sichuan Tit / 050

Sichuan Treecreeper / 114

Silver Oriole / 044

Slaty Bunting / 140

Snowy-cheeked Laughingthrush / 072

Sooty Bushtit / 054

Spectacled Fulvetta / 092

Spectacled Parrotbill / 098

Streaked Barwing / 064

Streaked Rosefinch / 130

T

Temminck's Tragopan / 020

Three-banded Rosefinch / 136

Three-toed Parrotbill / 096

Tibetan Babax / 080

W

White Eared Pheasant / 028

White-collared Yuhina / 106

White-speckled Laughingthrush / 076

Wood Snipe / 040

Y

Yunnan Nuthatch / 108

拉丁学名索引

A

Actinodura souliei / 064

Aegithalos fuliginosus / 054

Arborophila rufipectus / 016

Aythya baeri / 004

B

Bambusicola thoracicus / 018

C

Calliope obscura / 124

Calliope pectardens / 126

Carpodacus dubius / 138

Carpodacus rubicilloides / 130

Carpodacus trifasciatus / 136

Carpodacus verreauxii / 134

Carpodacus waltoni / 132

Certhia tianquanensis / 114

Cholornis paradoxus / 096

Chrysolophus amherstiae / 034

Chrysolophus pictus / 032

Crossoptilon auritum / 030

Crossoptilon crossoptilon / 028

E

Emberiza siemsseni / 140

F

Fulvetta cinereiceps / 094

Fulvetta ruficapilla / 092

Fulvetta striaticollis / 090

G

Gallinago nemoricola / 040

Garrulax canorus / 070

Grus nigricollis / 038

I

Ianthocincla bieti / 076

Ianthocincla lunulata / 078
Ianthocincla maximus / 074
Ianthocincla sukatschewi / 072

L

Larvivora ruficeps / 120
Liocichla omeiensis / 068
Locustella chengi / 058
Lophophorus lhuysii / 026
Lophophorus sclateri / 022

M

Mergus squamatus / 008
Moupinia poecilotis / 088
Myzornis pyrrhoura / 084

O

Oriolus mellianus / 044

P

Patayuhina diademata / 106
Perisoreus internigrans / 046
Phylloscopus emeiensis / 056
Poecile davidi / 048
Poecile weigoldicus / 050
Prunella immaculata / 128

Psittacula derbiana / 042
Pterorhinus berthemyi / 082
Pterorhinus koslowi / 080

S

Schoeniparus variegaticeps / 060
Sinosuthora conspicillata / 098
Sinosuthora przewalskii / 102
Sinosuthora zappeyi / 100
Sitta magna / 112
Sitta przewalskii / 110
Sitta yunnanensis / 108
Spizixos semitorques / 052
Suthora fulvifrons / 104

T

Tetraophasis obscurus / 012
Tetraophasis szechenyii / 014
Tetrastes sewerzowi / 010
Tragopan temminckii / 020
Trochalopteron elliotii / 066
Trochalopteron formosum / 062
Turdus kessleri / 116
Turdus mupinensis / 118

附录三　山地六条

影像设备日新月异，自然探索信息也借助互联网飞速传播，越来越多的朋友加入到生态摄影行列。

中国西南山地是国际公认的生物多样性热点区域，涉及川、滇、藏等多省。西南山地（成都山地文化传播有限公司）以地理冠名，创立十余年来，以"自然、科学、创作"为理念，用影像记录中国西南自然热点，运营中国西南最大的自然影像库www.swild.cn，创作生态影视作品，逐渐获得社会各界的广泛认可。

"博学之，审问之，慎思之，明辨之，笃行之。"《礼记·中庸》中的这一句话也应成为生态摄影的创作态度。作品是否优秀，取决于创作者的知识、思考和行为等层面。

以下六条（即"山地六条"），表明我们的态度和立场，愿与大家共识共勉。

第一条　中国西南山地区域仍存在大量生物多样性景观影像空白，影像拍摄是非常重要而真实的记录手段。

第二条　影像应正面推动科学信息传播、自然保护和社会可持续发展。

第三条　反对盲目追求"优美画面"而过度使用投食、播音和造景等手段。

第四条　自然声音和动态画面可以展现更加丰富的信息，鼓励录音和视频拍摄。

第五条　倡导科研工作者拍摄野外原生境物种影像，秉承科学态度，呈现高品质作品。

第六条　原创作品受法律保护，鼓励借助线上和线下平台传播，推动公益和商业的合法应用。

注："山地六条"表明西南山地SWILD（成都山地文化传播有限公司）的创作态度和运营立场，倡导自然影像行业良性发展，接受可行建议和监督。

图书在版编目（CIP）数据

中国西南山地珍稀特有鸟类：汉字、英文 / 西南山地SWILD编. -- 成都：四川美术出版社，2024.5
ISBN 978-7-5740-1135-9

Ⅰ.①中… Ⅱ.①西… Ⅲ.①山地－珍稀动物－鸟类－西南地区－普及读物－汉、英 Ⅳ.①Q959.708-64

中国国家版本馆CIP数据核字(2024)第090661号

中国西南山地
珍稀特有鸟类
ZHONGGUO XINAN SHANDI ZHENXI TEYOU NIAOLEI

西南山地SWILD⊙编

责任编辑	林雪红　袁一帆　余启敏
责任校对	陈　玲
责任印制	黎　伟
出版发行	四川美术出版社
地　　址	成都市锦江区工业园区三色路238号
成品尺寸	210mm×260mm
印　　张	11.5
字　　数	200千
图　　片	125幅
制　　作	成都华桐美术设计有限公司
印　　刷	成都市东辰印艺科技有限公司
版　　次	2024年11月第1版
印　　次	2024年11月第1次印刷
书　　号	ISBN 978-7-5740-1135-9
定　　价	129.00元

■ 著作权所有·侵权必究